**Biochemical Aspects
of Reactions on
Solid Supports**

Contributors

PEDRO CUATRECASAS
RACHEL GOLDMAN
LEON GOLDSTEIN
EPHRAIM KATCHALSKI
GARLAND R. MARSHALL
R. B. MERRIFIELD
JAMES A. PATTERSON
GEORGE R. STARK

Biochemical Aspects of Reactions on Solid Supports

Edited by
George R. Stark
Department of Biochemistry
Stanford University
School of Medicine
Stanford, California

1971 **Academic Press** New York and London

Copyright © 1971, by Academic Press, Inc.
ALL RIGHTS RESERVED
NO PART OF THIS BOOK MAY BE REPRODUCED IN ANY FORM,
BY PHOTOSTAT, MICROFILM, RETRIEVAL SYSTEM, OR ANY
OTHER MEANS, WITHOUT WRITTEN PERMISSION FROM
THE PUBLISHERS.

ACADEMIC PRESS, INC.
111 Fifth Avenue, New York, New York 10003

United Kingdom Edition published by
ACADEMIC PRESS, INC. (LONDON) LTD.
24/28 Oval Road, London NW1 7DD

LIBRARY OF CONGRESS CATALOG CARD NUMBER: 72-178217

PRINTED IN THE UNITED STATES OF AMERICA

Contents

	LIST OF CONTRIBUTORS	vii
	PREFACE	ix

Chapter 1. Water-Insoluble Enzyme Derivatives and Artificial Enzyme Membranes

Rachel Goldman, Leon Goldstein, and Ephraim Katchalski

I.	Introduction	1
II.	Preparation of Water-Insoluble Enzyme Derivatives	2
III.	Stability of Water-Insoluble Enzyme Derivatives	23
IV.	Kinetic Behavior of Particulate Immobilized Enzyme Systems	26
V.	Enzyme Columns	39
VI.	Enzyme Membranes	43
VII.	Applications	63
	References	72

Chapter 2. Selective Adsorbents Based on Biochemical Specificity

Pedro Cuatrecasas

I.	Introduction	79
II.	Water-Insoluble Carriers	82
III.	Selection of the Ligand	84
IV.	Covalent Linking Reactions	86
V.	Conditions for Chromatography on Affinity Columns	94
VI.	Importance of Anchoring Arms	95
VII.	Specific Adsorbents for Protein Purification	97
VIII.	Specific Nucleic Acid Adsorbents	104
IX.	Isolation of Cells, Receptor Structures, and Other Particulate Cell Structures	105
	References	107

Chapter 3. Solid Phase Synthesis: The Use of Solid Supports and Insoluble Reagents in Peptide Synthesis

Garland R. Marshall and R. B. Merrifield

I.	Introduction	111
II.	Polymeric Carboxyl Protection	115
III.	Polymeric Amino Protection	138

IV.	Polymeric Side-Chain Protection	141
V.	Polymeric Coupling Reagents	143
VI.	Effects of Physical Properties of Polymeric Supports	148
VII.	Automation	155
VIII.	Similar Development in Conventional Synthesis	156
IX.	Difficulties in Analysis	158
X.	Results	160
XI.	Conclusions and Summary	162
	References	162

Chapter 4. Sequential Degradation of Peptides Using Solid Supports
George R. Stark

I.	Introduction	171
II.	Degradation of Peptides Attached to Solid Supports	173
III.	Degradation of Peptides with an Insoluble Reagent	179
IV.	Properties of Resins Important for Peptide Degradations	184
	References	187

Chapter 5. Preparation of Cross-Linked Polystyrenes and Their Derivatives for Use as Solid Supports or Insoluble Reagents
James A. Patterson

I.	Introduction	189
II.	Types of Polystyrene and Their Properties	190
III.	The Preparation of Spherical Styrene-DVB Copolymers	195
IV.	Chemical Reactions of Styrene-Divinylbenzene Copolymers	201
	References	213
	AUTHOR INDEX	215
	SUBJECT INDEX	227

LIST OF CONTRIBUTORS

Numbers in parentheses indicate the pages on which the authors' contributions begin.

Pedro Cuatrecasas* (79), Laboratory of Chemical Biology, National Institute of Arthritis and Metabolic Diseases, National Institutes of Health, Bethesda, Maryland.

Rachel Goldman (1), Department of Biophysics, The Weizmann Institute of Science, Rehovot, Israel.

Leon Goldstein (1), Department of Biophysics, The Weizmann Institute of Science, Rehovot, Israel.

Ephraim Katchalski (1), Department of Biophysics, The Weizmann Institute of Science, Rehovot, Israel.

Garland R. Marshall (111), Departments of Physiology and Biophysics and of Biological Chemistry, Washington University Medical School, St. Louis, Missouri.

R. B. Merrifield (111), The Rockefeller University, New York, New York.

James A. Patterson (189), Sondell Scientific Instruments, Inc., San Antonio Road, Palo Alto, California.

George R. Stark (171), Department of Biochemistry, Stanford University School of Medicine, Stanford, California.

---

* Present address: Department of Medicine and Pharmacology, The Johns Hopkins University School of Medicine, Baltimore, Maryland.

Preface

A reactant or reagent attached covalently to an insoluble support can be separated rapidly and completely from a solution along with anything which has become attached to it either covalently or by adsorption. This property, perhaps somewhat unimpressive at first glance, can transform difficult separations and purifications into simple filtrations. The impact of the use of solid supports on biochemistry and chemistry is already appreciable and will surely grow. We hope that our presentation of the current state of the art will stimulate extension of the specific applications we have described and, more importantly, that our discussion of the general properties of polymers will be of use to those who seek to develop new applications of reactions on solid supports.

It will soon be apparent to the reader that each of the specific areas of application discussed in this book is in a different stage of development; therefore, some of the articles have substantial amounts of information in the form of unpublished data and personal communications. Such a situation is inevitable when an effort is made to cover the material in a rapidly evolving field in an up-to-date fashion.

There are some successful applications of solid state technology which are not discussed in this book, e.g., the uses of polymeric supports for synthesis and degradation of nucleotides. References to some recent articles on these subjects are provided at the end of this preface.

The technology which enables one to pluck a single enzyme specifically from a complex solution by adsorption to a solid support or to synthesize in large amounts a polypeptide which has the activity of a rare hormone is nothing less than the beginning of a revolution in biochemistry. This revolution has progressed far enough so that one can now sense its scale and even begin to appreciate what changes the full impact of the new order may bring.

References for Synthesis of Nucleotides on Polymeric Supports

1. Letsinger, R. L., and Mahadevan, V. (1965). *J. Amer. Chem. Soc.* **87**, 3526; *ibid.* (1966). **88**, 5319; Letsinger, R. L., Caruthers, M. H., and Jerina, D. M. (1967). *Biochemistry* **6**, 1379; Letsinger, R. L., Caruthers, M. H., Miller, P. S., and Ogilvie, K. K. (1967). *J. Amer. Chem. Soc.* **89**, 7146; Shimidzu, T., and Letsinger, R. L. (1968). *J. Org. Chem.*, p. 708.
2. Cramer, F., Helbig, R., Hettler, H., Scheit, K. H., and Seliger, H. (1966). *Angew. Chem. Int. Ed. Engl.* **5**, 601; Cramer, F., and Köster, H. (1968). *Angew. Chem. Int. Ed. Engl.* **7**, 473; Cramer, F. (1969). *Pure Appl. Chem.* **18**, 197; Cramer, F. (1969). *Colloq. CNRS* **182**, 343; Freist, W., and Cramer, F. (1970). *Angew. Chem. Int. Ed. Engl.* **9**, 368.
3. Hayatsu, H., and Khorana, H. G. (1966). *J. Amer. Chem. Soc.* **88**, 3162; *ibid.* (1967). **89**, 3880.
4. Blackburn, G. M., Brown, M. J., and Harris, M. R. (1967). *J. Chem. Soc. C*, p. 2438; Blackburn, G. M., Brown, M. J., Harris, M. R., and Shire, D. (1969). *J. Chem. Soc. C*, p. 676.
5. Melby, L. R., and Strobach, D. R. (1967). *J. Amer. Chem. Soc.* **89**, 450; Melby, L. R., and Strobach, D. R. (1969). *J. Org. Chem.*, p. 421; *ibid*, p. 427.

Reference for Degradation of Nucleotides

Wagner, T. E., Chai, H. G., and Warfield, A. S. (1969). *J. Amer. Chem. Soc.* **91**, 2388.

GEORGE R. STARK

**Biochemical Aspects
of Reactions on
Solid Supports**

CHAPTER 1 Water-Insoluble Enzyme Derivatives and Artificial Enzyme Membranes

RACHEL GOLDMAN, LEON GOLDSTEIN,
AND EPHRAIM KATCHALSKI

I.	Introduction		1
II.	Preparation of Water-Insoluble Enzyme Derivatives		2
	A.	Enzymes Immobilized by Adsorption	3
	B.	Enzymes Immobilized by Occlusion in Cross-Linked Polymeric Matrices	13
	C.	Enzymes Immobilized by Covalent Binding to Water-Insoluble Carriers	14
	D.	Enzymes Immobilized by Intermolecular Cross-Linking	22
III.	Stability of Water-Insoluble Enzyme Derivatives		23
	A.	Storage Stability	23
	B.	Thermal Stability	24
	C.	Dependence of Stability on pH	25
IV.	Kinetic Behavior of Particulate Immobilized Enzyme Systems		26
	A.	Effects of Diffusion Limitations	26
	B.	Steric Effects	27
	C.	Effects of Chemical Modification	28
	D.	Microenvironmental Effects	29
	E.	Immobilized Multienzyme Systems	38
V.	Enzyme Columns		39
VI.	Enzyme Membranes		43
	A.	Structure of Enzyme-Collodion Membranes	43
	B.	Dependence of Membrane Enzymic Activity on pH	45
	C.	Analysis of the Kinetic Behavior of Enzyme Membranes	51
VII.	Applications		63
	References		72

I. INTRODUCTION

The main interest in water-insoluble enzyme derivatives stems from their possible use as heterogeneous specific catalysts in research and industry. Furthermore, since many enzymes are embedded in biological membranes and subcellular particles, artificially immobilized enzymes can serve as model systems for the investigation of the effect of microenvironment on the mode of action of enzymes.

A considerable number of immobilized enzyme systems have been described in the literature. Of particular interest are systems consisting of enzymes immobilized within cross-linked polymeric networks, enzymes covalently bound to water-insoluble high molecular weight carriers and enzymes embedded within artificial membranes.

Water-insoluble enzyme derivatives provide specific, easily removable reagents. If stable, they can be used repeatedly. Enzyme columns and enzyme membranes can be used in continuous enzymic processes.

In the following, the preparation of the different types of immobilized enzyme systems is given. The various factors affecting the properties and the apparent kinetic behavior of immobilized enzymes are discussed. Current developments and trends in the application of water-insoluble enzyme derivatives are surveyed.

Several review articles dealing with water-insoluble enzyme derivatives are available (Manecke, 1964; Silman and Katchalski, 1966; Sehon, 1967; Goldstein and Katchalski, 1968; Goldstein, 1969).

II. PREPARATION OF WATER-INSOLUBLE ENZYME DERIVATIVES

Four principal methods have been used for the preparation of water-insoluble derivatives of biologically active proteins (Manecke, 1964; Silman and Katchalski, 1966; Sehon, 1967; Goldstein and Katchalski, 1968; Goldstein, 1969): (a) adsorption on inert carriers or synthetic ion exchange resins; (b) occlusion into gel lattices, the pores of which are too small to allow the escape of the entrapped protein; (c) covalent binding of proteins to a suitable water-insoluble carrier, via functional groups not essential for their biological activity; and (d) covalent cross-linking of the protein by an appropriate bifunctional reagent. Immobilized enzyme derivatives have also been prepared by adsorbing the protein onto a suitable support material followed by intermolecular cross-linking of the enzyme.

The methods available for the immobilization of proteins have been recently summarized in several reviews (Manecke, 1964; Silman and Katchalski, 1966; Sehon, 1967; Goldstein and Katchalski, 1968; Goldstein, 1969). Some of the more important methods are listed below.

A. Enzymes Immobilized by Adsorption

The adsorption of proteins by different adsorbents might often lead to denaturation. A suitable adsorbent should thus possess high affinity for the enzyme protein and yet cause minimal denaturation. Proteins are known to adsorb nonspecifically onto both charged resins and neutral surfaces (see Table IA). Both types of solid supports have been used for the immobilization of enzymes. High concentrations of salt or substrate enhance the rate of desorption of proteins. Adsorption techniques are thus of limited reliability when complete enzyme insolubilization is desired. Desorption of enzymes from solid surfaces can be overcome, as previously mentioned, by cross-linking the protein subsequent to its adsorption.

The early work on the adsorption of proteins on various surfaces has been summarized in two reviews (Zittle, 1953; Silman and Katchalski, 1966).

Several ion-exchange resins such as DEAE-cellulose, DEAE-Sephadex, and CM-cellulose have been used as solid supports for the immobilization of enzymes by ionic bonding. To cite some examples: Nikolayev and Mardashev (1961; Nikolayev, 1962) prepared complexes of asparaginase with CM-cellulose and with DEAE-cellulose. DEAE-cellulose has also been employed for the immobilization of invertase (Suzuki et al., 1966). A fungal amino acylase adsorbed on DEAE-cellulose and DEAE-Sephadex has been recently used in column form for the continuous resolution of racemic mixtures of N-acetyl amino acids on an industrial scale (Tosa et al., 1966, 1967a,b, 1969a,b).

Adsorption of enzymes onto glass beads, quartz, and charcoal particles, dialysis tubing, Millipore filters, etc., has been described (Mitz, 1956; Kobamoto et al., 1966; Poltorak and Vorobeva, 1966; Goldfeld et al., 1966; Vorobeva and Poltorak, 1966a,b). Enzyme inactivation has been observed in some cases (Poltorak and Vorobeva, 1966; Goldfeld et al., 1966; Vorobeva and Poltorak, 1966a,b). However, immobilized enzyme systems of high activity have been prepared by utilizing the high adsorption capacity for protein of silica gel, charcoal and quartz, as well as collodion membranes (Poltorak and Vorobeva, 1966; Goldfeld et al., 1966; Vorobeva and Poltorak, 1966a,b; Goldman et al., 1965, 1968a; Goldman and Katchalski, 1971; Haynes and Walsh, 1969). Studies of the mechanism of adsorption of papain (Goldman et al., 1965, 1968) and alkaline phosphatase (Goldman and Katchalski, 1971) onto collodion membranes, and of trypsin onto silica gel particles (Haynes and Walsh, 1969), have indicated that these proteins form a monomolecular layer on the adsorbing surface. Fixation of the adsorbed protein could be accomplished by intermolecular cross-linking by means of bisdiazobenzidine-2,2'-disulfonic acid in the case of the papain-collodion conjugate (Goldman et al., 1965, 1968a) and by glutaraldehyde in the case of the trypsin-silica gel (Haynes and Walsh, 1969) and alkaline phosphatase collodion conjugates (Goldman and Katchalski, 1971).

TABLE I

IMMOBILIZED ENZYMES

A. ENZYMES IMMOBILIZED BY ADSORPTION

Enzyme	Adsorbent	Reference
Acid phosphatase	Silica gel	Vorobeva and Poltorak (1966b)
	Charcoal	Vorobeva and Poltorak (1966b)
Acylase	DEAE–cellulose	Tosa et al. (1966, 1967a, b, 1969a); (Chibata et al. 1966a,b)
	DEAE–Sephadex	Tosa et al. (1966, 1967a, b, 1969a,b); Chibata et al. (1966a,b)
Alkaline phosphatase	Collodion membrane (followed by cross-linking with glutaraldehyde)	Goldman and Katchalski (1969, 1971)
	Silica gel	Poltorak and Vorobeva (1966)
	Charcoal	Poltorak and Vorobeva (1966)
Asparaginase	CM–cellulose	Nikolayev and Mardashev (1961)
	DEAE–cellulose	Nikolayev (1962)
Catalase	Cellulose anion exchanger	Mitz and Yanari (1964)
	Silica gel	Goldfeld et al. (1966)
	Charcoal	Goldfeld et al. (1966)
Chymotrypsin	Carboxymethyl cellulose	Mitz and Schleuter (1959)
	Cellulose citrate	Mitz and Schleuter (1959)
	Cellulose phosphate	Mitz and Schleuter (1959)
	Kaolinite	McLaren (1957, 1960); McLaren and Babcock (1959); McLaren and Estermann (1956, 1957); McLaren et al. (1958)
Diastase (amylase)	Activated charcoal	Stone (1955)
	Acid clay	Usami and Taketomi (1965)
Glucoamylase	DEAE-cellulose	Bachler et al. (1970)
Glucose oxidase	Cellophane sheets (followed by cross-linking with a bisdiazobenzidine derivative or with glutaraldehyde)	Broun et al. (1969); Sélégny et al. (1968)

TABLE I (continued)

Enzyme	Adsorbent	Reference
Glucose-6-phosphate dehydrogenase	Collodion membrane	Goldman and Lenhoff (1969)
Hexokinase	Silica gel	Vorobeva and Poltorak (1966a)
	Charcoal	Vorobeva and Poltorak (1966a)
Invertase	DEAE–cellulose	Suzuki et al. (1966)
Lipase	Amberlite XE–97	Brandenberger (1956)
Papain	Collodion membrane (followed by cross-linking with bisdiazobenzidine-2,2′-disulfonic acid)	Goldman et al. (1965, 1968a,b)
Phosphoglucomutase	Silica gel	Vorobeva and Poltorak (1966b)
	Charcoal	Vorobeva and Poltorak (1966b)
Trypsin	Carboxymethyl cellulose	Mitz and Schleuter (1959)
	Cellulose citrate	Mitz and Schleuter (1959)
	Cellulose phosphate	Mitz and Schleuter (1959)
	Cellulose, glass, quartz, and dialysis tubing	Kobamoto et al. (1966)
	Silica gel (followed by cross-linking with glutaraldehyde)	Haynes and Walsh (1969)
Ribonuclease	Dowex-50	Barnett and Bull (1959)
	Glass	Hummel and Anderson (1965)

B. ENZYMES IMMOBILIZED BY OCCLUSION IN CROSS-LINKED POLYMERIC MATRICES

Enzyme	Polymer matrix	Reference
Acetylcholinesterase	Starch gel	Aldrich et al. (1963, 1965) Bauman et al. (1965, 1967) Guilbault and Kramer (1965)
	Silastic	Pennington et al. (1968)
Alcohol dehydrogenase	Polyacrylamide gel	Wieland et al. (1966)
Aldolase	Polyacrylamide gel	Bernfeld et al. (1968)
Cholinesterase	Polyacrylamide gel	Degani and Miron (1970)
	Polyacrylamide gel	Guilbault and Das (1970)
	Starch gel	Guilbault and Das (1970)
Chymotrypsin	Polyacrylamide gel	Bernfeld and Wan (1963)
Citrate synthetase	Polyacrylamide gel	Mosbach (1970)

TABLE I (continued)

Enzyme	Polymer matrix	Reference
Diastase (amylase)	Polyacrylamide gel	Bernfeld and Wan (1963)
Glucose oxidase	Polyacrylamide gel	Hicks and Updike (1966)
Glutamic-pyruvic transaminase	Polyamides	Leuschner (1964, 1966)
	Cellulose acetate	Leuschner (1964, 1966)
Lactic dehydrogenase	Polyacrylamide gel	Hicks and Updike (1966); Wieland et al. (1966)
Orsellinic decarboxylase	Polyacrylamide gel	Mosbach and Mosbach (1966)
Papain	Polyacrylamide gel	Bernfeld and Wan (1963)
Phosphoglycerate mutase	Polyacrylamide gel	Bernfeld et al. (1969a,b)
Steroid Δ^1-dehydrogenase	Polyacrylamide gel	Mosbach and Larsson (1970)
Trypsin	Polyacrylamide gel	Bernfeld and Wan (1963); Wieland et al. (1966)
Urease	Polyacrylamide gel	Guilbault and Montalvo (1969, 1970)
	Polyacrylamide gel	Guilbault and Das (1970)
	Microcapsules (nylon and collodion)	Chang (1964, 1966)

C. ENZYMES IMMOBILIZED BY COVALENT BINDING TO WATER-INSOLUBLE CARRIERS

Enzyme	Carrier (active form given in parentheses)	Reference
Acylase	Poly-α-amino acids (polypeptidylation, by initiating the polymerization of N-carboxy anhydride of amino acid with enzyme)	Kirimura and Yoshida (1966)
	Porous glass (aminoalkylsilane glass converted to the aminoaryl derivative)	Weetall (1969b)
Adenosine triphosphatase	Carboxymethyl cellulose (CMC-azide)	Brown et al. (1966, 1967)
Aldolase	Aminoethyl cellulose (coupled via glutaraldehyde)	Lynn and Falb (1969)
Alkaline phosphatase	Ethylene-maleic anhydride copolymer (EMA)	Zingaro and Uziel (1970)
	Carboxymethyl cellulose (CMC-azide)	Zingaro and Uziel (1970)
	Copolymer of methacrylic acid and methacrylic m-fluoroanilide	Zingaro and Uziel (1970)

TABLE I (continued)

Enzyme	Carrier (active form given in parentheses)	Reference
	Porous glass (amino-alkylsilane glass converted to the amino aryl derivative)	Weetall (1969)
Amino acyl t-RNA synthetase	Sepharose (cyanogen bromide activation)	Denburg and De Luca (1970)
Amylase	3-(p-Aminophenoxy)-2-hydroxypropyl ether of cellulose (polydiazonium salt)	Barker et al. (1968); Barker et al. (1969)
	2-Hydroxy-3-(p-isothio-cyanatophenoxy)propyl ether of cellulose	Barker et al. (1968); Barker et al. (1969)
	Polyacrylamide beads (through the acyl azide, isothiocyanato compound; or the polydiazonium salt)	Barker et al. (1970)
Amyloglucosidase	DEAE-cellulose (2-amino-1,6-dichloro-s-triazine activation)	Wilson and Lilly (1969)
Apyrase	Carboxymethyl cellulose	Whittam et al. (1968)
	Carboxymethyl cellulose, alanine-glutamic acid copolymer, polyaspartic acid, polygalacturonic acid (activation by Woodward's Reagent K)	Patel et al. (1969)
	Ethylene-maleic anhydride copolymer (EMA)	Patel et al. (1969)
	Carboxymethyl cellulose (CMC-azide)	Wheeler et al. (1969)
	Cellulose (activated by cyanuric chloride)	Wheeler et al. (1969)
Bromelain	Carboxymethyl cellulose (CMC-azide)	Wharton et al. (1968a)
Carboxypeptidase	Poly-p-aminostyrene (polydiazonium salt)	Grubhofer and Schleith (1953, 1954)
	Sepharose (cyanogen bromide activation)	Seki et al. (1970,b)
Catalase	Cellulose (cyanuric chloride activation)	Surinov and Manoylov (1966)
	m-Aminobenzyloxy methyl ether of cellulose (polydiazonium salt)	Surinov and Manoylov (1966)

TABLE I (continued)

Enzyme	Carrier (active form given in parentheses)	Reference
Cholinesterase	Carboxymethyl cellulose, alanine-glutamic acid copolymer; or polyaspartic acid (activation by Woodward's Reagent K)	Patel et al. (1969)
	Ethylene-maleic anhydride copolymer (EMA)	Patel et al. (1969)
Chymotrypsin	-Amino-DL-phenylalanine-L-leucine copolymer (polydiazonium salt)	Katchalski (1962)
	Ethylene-maleic anhydride copolymer (EMA)	Levin et al. (1964); Goldstein (1970); Fritz et al. (1966, 1967)
	p-Aminobenzyl cellulose (polydiazonium salt)	Mitz and Summaria (1961); Lilly et al. (1965)
	m-Aminobenzyloxy methyl ether of cellulose (polydiazonium salt)	Surinov and Manoylov (1966)
	Carboxymethyl cellulose (CMC-azide)	Mitz and Summaria (1961); Lilly et al. (1965)
	Poly-α-amino acids (polypeptidylation, by initiating the polymerization of N-carboxy anhydride of amino acid with enzyme)	Kirimura and Yoshida (1966)
	Cross-linked polysaccharides—Sephadex, Sepharose, Agarose (cyanogen bromide activation)	Axén et al. (1967, 1969, 1970); Porath et al. (1967); Porath (1967); Green and Crutchfield (1969)
	Sephadex (isocyanate derivative)	Axén and Porath (1966)
	Cellulose (cyanuric chloride activation)	Kay and Crook (1967); Surinov and Manoylov (1966)
	Cellulose, DEAE-cellulose, Sephadex, or Sepharose (activated by 2-amino-4,6-dichloro-s-triazine)	Kay and Lilly (1970)
	Alanine-glutamic acid (10 : 1) copolymer (activation by Woodward's Reagent K)	Wagner et al. (1968)
	Polyglutamic acid, carboxymethyl cellulose, polyacrylic acid (activation by Woodward's Reagent K)	Patel et al. (1967)

TABLE I (continued)

Enzyme	Carrier (active form given in parentheses)	Reference
Creatine kinase	DEAE-cellulose paper (cyanuric chloride activation)	Kay *et al.* (1968)
Deoxyribonuclease	Carboxymethyl cellulose, alanine-glutamic acid copolymer, polyaspartic acid, polygalacturonic acid (activated by Woodward's Reagent K)	Patel *et al.* (1969)
	Ethylene-maleic anhydride copolymer (EMA)	Patel *et al.* (1969)
Diastase (amylase)	Poly-*p*-aminostyrene (polydiazonium salt)	Grubhofer and Schleith (1953, 1954)
	Copolymer of methacrylic acid and methacrylic *m*-fluoroanilide	Manecke and Günzel (1967); Manecke and Singer (1960)
	Nitrofluorobenzene sulfonyl chloride derivative of polystyrene	Manecke and Förster (1966)
Ficin	Carboxymethyl cellulose (CMC-azide)	Lilly *et al.* (1965)
Fructose-1, 6-diphosphatase	Aminoethyl cellulose (coupled via glutaraldehyde)	Lynn and Falb (1969)
β-Galactosidase	Cellulose (cyanuric chloride activation)	Kay *et al.* (1968)
	DEAE-cellulose paper (cyanuric chloride activation)	Kay *et al.* (1968)
Glucose-6-phosphate dehydrogenase	Separose (cyanogen bromide activation)	Mosbach and Mattiasson (1970)
	Acrylamide-acrylic acid copolymer (carbodiimide activation)	Mosbach and Mattiasson (1970)
Glucose oxidase	Polystyrene tubing	Hornby *et al.* (1970)
	Nickel oxide on nickel screens	Weetall and Hersh (1970)
Glyceraldehyde-3-phosphate dehydrogenase	Aminoethyl cellulose (coupled via glutaraldehyde)	Lynn and Falb (1969)
Hexokinase	Sepharose (cyanogen bromide activation)	Mosbach and Mattiasson (1970)
	Acrylamide-acrylic acid copolymer (carbodiimide activation)	Mosbach and Mattiasson (1970)
Invertase	Copolymer of methacrylic acid and methacrylic-*m*-fluoroanilide	Manecke and Singer (1960); Manecke (1962)

TABLE I (continued)

Enzyme	Carrier (active form given in parentheses)	Reference
Lactic dehydrogenase	DE-81 cellulose anion-exchange paper (activated by s-dichlorotriazinyl dye: "Procion brilliant orange MGS")	Wilson et al. (1968a)
Lipase	Poly-p-aminostyrene (polydiazonium salt)	Brandenberger (1956)
	Carboxy chloride resins	Brandenberger (1956)
Luciferase	Polyacrylic acid (through the acyl azide)	Erlanger et al. (1970)
Papain	p-Amino-DL-phenylalanine-L-leucine copolymer (polydiazonium salt)	Silman et al. (1966); Cebra et al. (1961)
	p-Aminobenzyl cellulose (polydiazonium salt)	Goldstein et al. (1970); Kominz et al. (1965)
	S-MDA Resin [polydiazonium salt (see Fig. 7)]	Goldstein (1970); Goldstein et al. (1970)
	Ethylene-maleic anhydride copolymer (EMA)	Goldstein (1970)
	Nitrofluorobenzene sulfonyl chloride derivative of polystyrene	Manecke and Förster (1966)
	Porous glass (aminoalkylsilane glass converted to the isothiocyanate derivative)	Weetall (1969a)
	Porous glass (aminoalkylsilane glass converted to the amino aryl derivative)	Weetall (1969a)
Pepsin	Poly-p-aminostyrene (polydiazonium salt)	Grubhofer and Schleith (1953)
	p-Amino-DL-phenylalanine-L-leucine copolymer (polydiazonium salt)	Katchalski (1962)
	Methacrylic acid-methacrylic acid-3-fluoro-4,6-dinitroanilide	Manecke (1962)
Peroxidase	Carboxymethyl cellulose (activated with carbodiimides)	Weliky et al. (1969)
Pronase	Bromoacetyl cellulose	Shaltiel et al. (1970)
	Carboxymethyl Sephadex (carbodiimide activation)	Shaltiel et al. (1970)
	p-Amino-DL-phenylalanine-L-leucine copolymer (polydiazonium salt)	Cresswell and Sanderson (1970)
Proteinase (bacterial)	Agarose (cyanogen bromide activation)	Gabel and Hofsten (1970)

TABLE I (continued)

Enzyme	Carrier (active form given in parentheses)	Reference
Prothrombin	p-Amino-DL-phenylalanine-L-leucine copolymer (polydiazonium salt)	Engel and Alexander (1971)
Pyruvate kinase	Filter paper (cellulose-activated with cyanuric chloride)	Wilson et al. (1968b)
Rennin	Agarose (activated with cyanogen bromide)	Green and Crutchfield (1969)
	Aminoethyl cellulose (coupled with glutaraldehyde)	Green and Crutchfield (1969)
Renin	Sepharose (cyanogen bromide activation)	Seki et al. (1970a)
Ribonuclease	Carboxymethyl cellulose (CMC-azide)	Lilly et al. (1965); Epstein and Anfinsen (1962)
	Cellulose (activated with cyanuric chloride)	Surinov and Manoylov (1966)
	m-Aminobenzyloxy methyl ether of cellulose (polydiazonium salt)	Surinov and Manoylov (1966)
	p-Aminobenzyl cellulose (polydiazonium salt)	Lilly et al. (1965)
	p-Amino-DL-phenylalanine-L-leucine copolymer (polydiazonium salt)	Silman et al. (1963)
Streptokinase	p-Aminobenzyl cellulose (polydiazonium salt)	Steinbuch and Pejaudier (1964)
	p-Amino-DL-phenylalanine-L-leucine copolymer (polydiazonium salt)	A. Rimon et al. (1963); Gutman and Rimon (1964)
	Ethylene-maleic anhydride copolymer (EMA)	S. Rimon et al. (1966)
Subtilopeptidase A (subtilisin Carlsberg)	Ethylene-maleic anhydride copolymer (EMA)	Goldstein (1970)
	S-MDA resin [polydiazonium salt (see Fig. 7)]	Goldstein (1970); Goldstein et al. (1970)
Thrombin	Bromoacetyl cellulose	Newcomb and Hoshida (1965)
	p-Amino-DL-phenylalanine-L-leucine copolymer (polydiazonium salt)	Hussain and Newcomb (1964)
	Ethylene-maleic anhydride copolymer (EMA)	Engel and Alexander (1966); Cohen et al. (1966)
Trypsin	Carboxymethyl cellulose (CMC-azide)	Mitz and Summaria (1961); Epstein and Anfinsen (1962)
	Ethylene-maleic anhydride copolymer (EMA)	Levin et al. (1964); Alexander et al. (1965, 1966); Fritz et al. (1966, 1969)

TABLE I (continued)

Enzyme	Carrier (active form given in parentheses)	Reference
	p-Amino-DL-phenylalanine-L-leucine copolymer (polydiazonium salt)	Bar Eli and Katchalski (1963); Bar Eli and Katchalski (1960)
	Bromoacetyl cellulose	Patchornik (1962)
	S-MDA Resin [polydiazonium salt (see Fig. 7)]	Goldstein (1970); Goldstein et al. (1970)
	Sephadex (isocyanate derivative)	Axén and Porath (1966)
	Porous glass (aminoalkylsilane glass converted to the isothiocyanate derivative)	Weetall (1969a)
	Porous glass (aminoalkylsilane glass converted to the amino aryl derivative)	Weetall (1969a)
	Polyacrylamide beads (through the acyl azide)	Inman and Dintzis (1969)
	Aminoethyl cellulose (coupled via glutaraldehyde)	Glassmeyer and Ogle (1971)
	Sephadex, Sepharose (cyanogen bromide activation)	Gabel et al. (1970)
	Cellulose	Craven and Gupta (1970)
	Acrylamide-acrylic acid copolymer (carbodiimide activation)	Mosbach (1970)
	Acrylamide-hydroxyethyl methacrylate copolymer (cyanogen bromide) activation)	Mosbach (1970)
	Nylon tubes	Hornby and Filippuson (1970)
	Alanine-glutamic acid (10 : 1) copolymer (activated by Woodward's Reagent K)	Wagner et al. (1968)
Urease	p-Amino-DL-phenylalanine-L-leucine copolymer (polydiazonium salt)	Riesel and Katchalski (1964)
	Nylon tubes	Sundaram and Hornby (1970)
	Porous glass (aminoalkylsilane glass converted to the amino aryl derivative)	Weetall and Hersh (1969)

TABLE I (continued)

D. ENZYMES IMMOBILIZED BY INTERMOLECULAR CROSS-LINKING

Enzyme	Cross-linking reagent	Reference
Carboxypeptidase	Glutaraldehyde	Quiocho and Richards (1964, 1966a,b)
Catalase	Glutaraldehyde	Schejter and Bar-Eli (1970)
Chymotrypsin	N-Ethyl-5-phenylisoxazolium-3′-sulfonate (Woodward's Reagent K)	Patel and Price (1961)
Papain	Bisdiazobenzidine 2,2′-disulfonic acid	Silman et al. (1966)
	Diphenyl-4,4′-diisothiocyanate 2,2′-disulfonic acid	Manecke and Günzel (1967)
	Glutaraldehyde	Jansen and Olson (1969); Ashoor et al. (1971); Ottesen and Svensson (1971)
Ribonuclease	Dimethyl adipimidate	Hartman and Wold (1967)
	Glutaraldehyde	Richards (1963)
	1,5-Difluoro-2,4-dinitrobenzene	Manfrey and King (1965)
Subtilopeptidase B (subtilisin Novo)	Glutaraldehyde	Ogata et al. (1968)
Trypsin	Glutaraldehyde	Habeeb (1967)

B. Enzymes Immobilized by Occlusion in Cross-Linked Polymeric Matrices

Enzymes can be occluded within a cross-linked gel matrix carrying out the polymerization reaction leading to gel formation in an aqueous solution containing the enzyme. In most of the cases reported (see Table IB) acrylamide was used as monomer and cross-linking was effected by means of N,N-methylene bis(acrylamide) (Bernfeld and Wan, 1963; Arnold, 1966; Hicks and Updike, 1966; Wieland et al., 1966; Penzer and Radda, 1967; van Duijn et al., 1967; Bernfeld et al., 1969b). The resulting block of polymerized enzyme gel can be mechanically dispersed into particles of defined size which might be lyophilized (Hicks and Updike, 1966).

The entrapping method imposes minimal constraints on the immobilized enzyme and does not involve covalent-bond formation with the supporting matrix. It might, therefore, in principle be applied to any enzyme. Several intrinsic drawbacks of the method should, however, be pointed out. (a) Because of the broad distribution in the pore size of synthetic gels of the polyacrylamide type, continuous leakage of the occluded enzyme can hardly be avoided. (b) The enzymic reaction occurs only within the domain of the gel matrix. The catalytic reaction is thus limited to substrates which can diffuse readily into the gel. (c) The free radicals generated in the course of polymerization might markedly effect the activity of the entrapped enzyme.

A considerable number of enzymes, such as glucose oxidase (Hicks and Updike, 1966), lactic dehydrogenase (Hicks and Updike, 1966; Wieland *et al.*, 1966), trypsin (Bernfeld and Wan, 1963; Wieland *et al.*, 1966), α-amylase (Bernfeld and Wan, 1963), and aldolase (Bernfeld and Wan, 1963; Bernfeld *et al.*, 1968) have been entrapped in acrylamide gels. Acetylcholinesterase has been immobilized by inclusion in starch gel-stabilized onto polyurethane-foam pads (Aldrich *et al.*, 1963, 1965; Guilbault and Kramer, 1965; Bauman *et al.*, 1965). The potential of the above preparations in biochemical analysis has been illustrated (see Section VII). The use of a silicon polymer, silastic, as a cross-linked network for entrapping cholinesterase has been recently reported (Pennington *et al.*, 1968). The silastic gel enhanced the thermal stability of the enzyme; the permeability of the gel to substrates and inhibitors seems, however, to be low.

Microcapsules, made of thin spherical semipermeable nylon or collodion membranes, have been used to enclose various enzymes (Chang, 1964, 1966; Chang *et al.*, 1966). Several microencapsulated enzymes, such as carbonic anhydrase, trypsin, and uricase, were shown to retain their catalytic activity on low molecular weight substrates (Chang, 1964, 1966; Chang *et al.*, 1966).

C. Enzymes Immobilized by Covalent Binding to Water-Insoluble Carriers

Binding of enzymes to insoluble carriers by covalent bonds should be carried out via functional groups on the protein which are not essential for its catalytic activity. The binding reaction should obviously be performed under conditions which do not cause denaturation. In determining the nature of the covalent bonds by which a given protein should be insolubilized, and selecting the appropriate water-insoluble carrier, use can be made of the information available on the effects of chemical modification of proteins on their biological activity (see, for example, reviews by Fraenkel-Conrat, 1959; Sri Ram *et al.*, 1962; Vallee and Riordan, 1969). The functional groups of proteins suitable for covalent binding under mild conditions include the following: (1) α- and ε-amino groups; (2) α-, β-, and γ-carboxyl groups; (3) the phenol ring of tyrosine; (4) the sulfhydryl and hydroxyl groups of cysteine and serine, respectively, and (5) the imidazole group of histidine. Only the first three classes of functional groups have been extensively utilized in the preparation of water-insoluble enzyme derivatives.

Enzymes have been covalently linked to carboxylic polymers via their amino groups (Manecke, 1964; Silman and Katchalski, 1966; Sehon, 1967; Goldstein

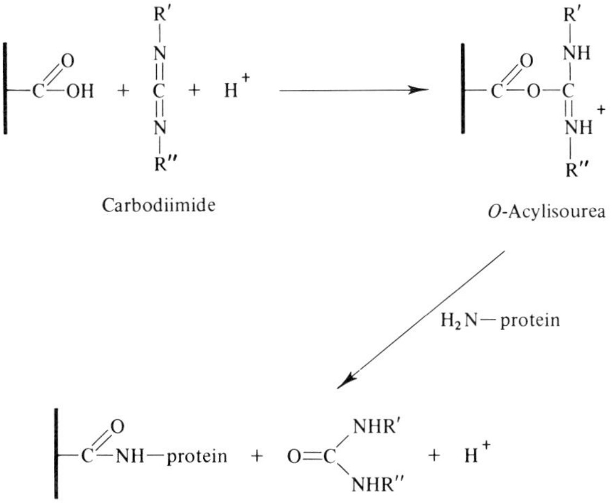

Fig. 1. Coupling of proteins to carboxylic polymers by the carbodiimide method. [Weliky and Weetall (1965).]

and Katchalski, 1968; Goldstein, 1969). The carrier carboxyls were activated by means of carbodiimides (Weliky and Weetall, 1965; Hoave and Koshland, 1967; Weliky *et al.*, 1969) (Fig. 1), by Woodward's Reagent K (*N*-ethyl-5-phenylisoxazolium-3'-sulfonate) (Patel *et al.*, 1967), or by their transformation into the corresponding azides (Micheel and Evers, 1949; Mitz and Summaria, 1961; Hornby *et al.*, 1966; Wharton *et al.*, 1968a; Whittam *et al.*, 1968) (Fig. 2).

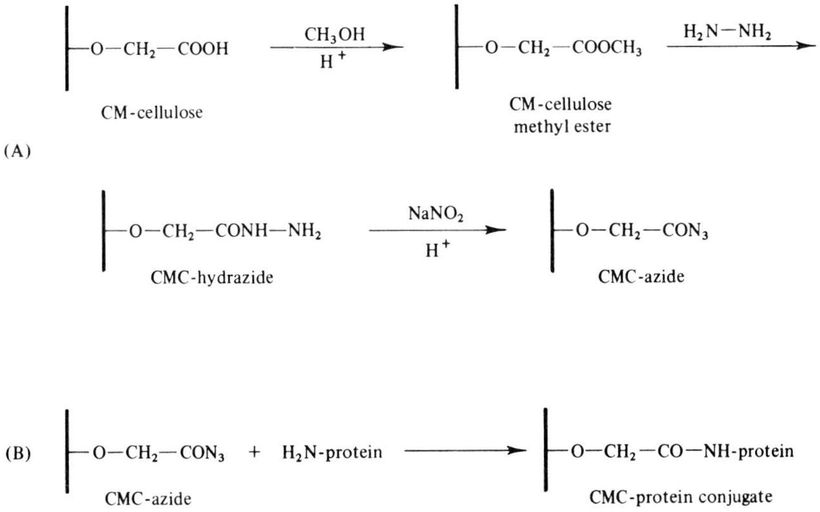

Fig. 2. Synthesis of carboxymethyl cellulose azide (A) and of carboxymethyl cellulose-protein conjugates (B). [Micheel and Evers (1949)].

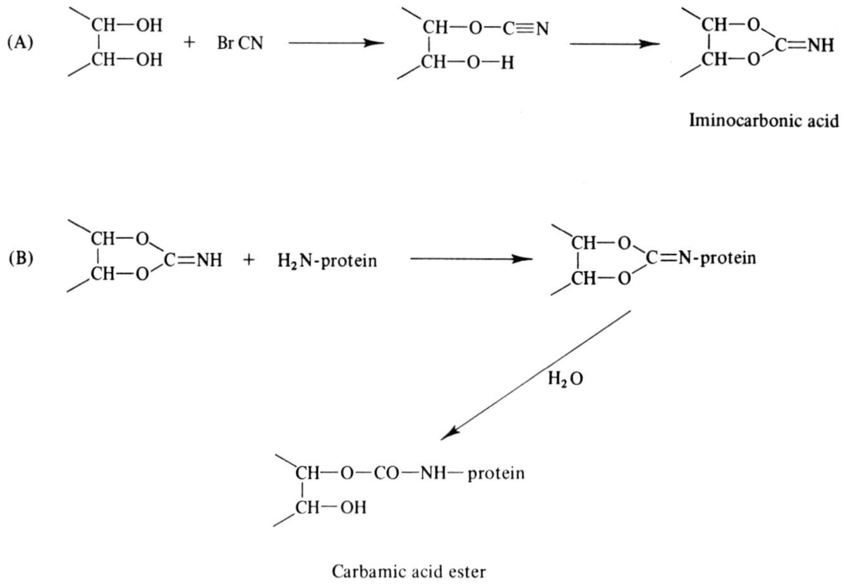

(A) $\begin{array}{c}\diagdown\text{CH}-\text{OH}\\ |\\ \diagup\text{CH}-\text{OH}\end{array}$ + BrCN $\longrightarrow$ $\begin{array}{c}\diagdown\text{CH}-\text{O}-\text{C}\equiv\text{N}\\ |\\ \diagup\text{CH}-\text{O}-\text{H}\end{array}$ $\longrightarrow$ $\begin{array}{c}\diagdown\text{CH}-\text{O}\diagdown\\ |\quad\quad\quad\text{C}=\text{NH}\\ \diagup\text{CH}-\text{O}\diagup\end{array}$

Iminocarbonic acid

(B) $\begin{array}{c}\diagdown\text{CH}-\text{O}\diagdown\\ |\quad\quad\quad\text{C}=\text{NH}\\ \diagup\text{CH}-\text{O}\diagup\end{array}$ + H_2N-protein $\longrightarrow$ $\begin{array}{c}\diagdown\text{CH}-\text{O}\diagdown\\ |\quad\quad\quad\text{C}=\text{N-protein}\\ \diagup\text{CH}-\text{O}\diagup\end{array}$

$\downarrow H_2O$

$\begin{array}{c}\diagdown\text{CH}-\text{O}-\text{CO}-\text{NH}-\text{protein}\\ |\\ \diagup\text{CH}-\text{OH}\end{array}$

Carbamic acid ester

Fig. 3. Coupling of proteins to Sepharose by means of cyanogen bromide. [Porath (1967).]

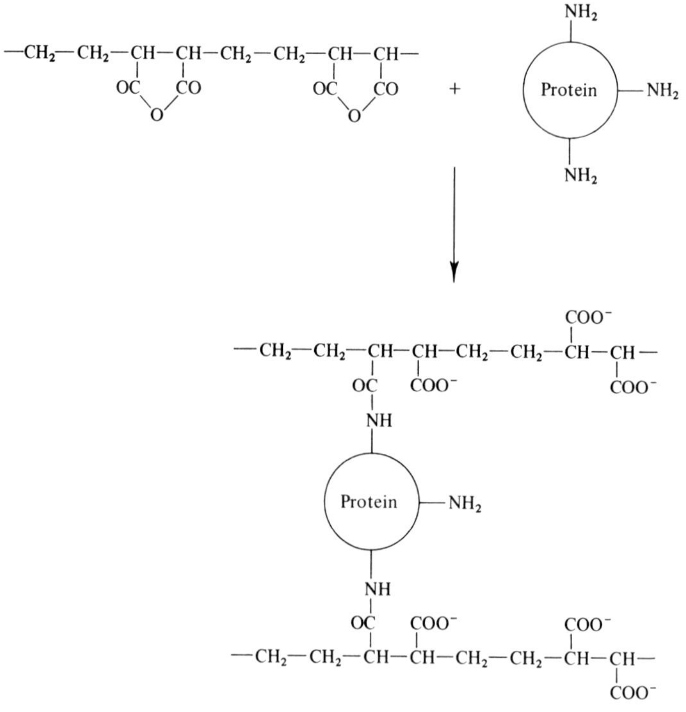

Fig. 4. Coupling of proteins to a copolymer of ethylene and maleic anhydride (EMA). [Levin et al. (1964).]

More recently several enzymes have been bound, via their amino groups, to cellulose activated by *sym*-trichlorotriazine (cyanuric chloride) (Surinov and Manoylov, 1966; Kay and Crook, 1967; Kay et al., 1968; Wilson et al., 1968a) or by a dichloro-*sym*-triazinyl dyestuff (Procion brilliant orange MGS) (Wilson et al., 1968a), as well as to Sephadex or Sepharose, activated by cyanogen bromide (Porath, 1967; Porath et al., 1967; Axén et al., 1967, 1969, 1971) (Fig. 3). A polymeric acylating reagent, ethylene-maleic anhydride (1:1) copolymer (EMA), has been successfully used for the preparation of polyanionic water-insoluble derivatives of enzymes, antigens and protein enzyme inhibitors (Fig. 4) (Levin et al., 1964; Alexander et al., 1965, 1966; Silman and Katchalski, 1966; Fritz et al., 1966, 1967, 1968, 1969; Sehon 1967; Goldstein and Katchalski, 1968; Goldstein,

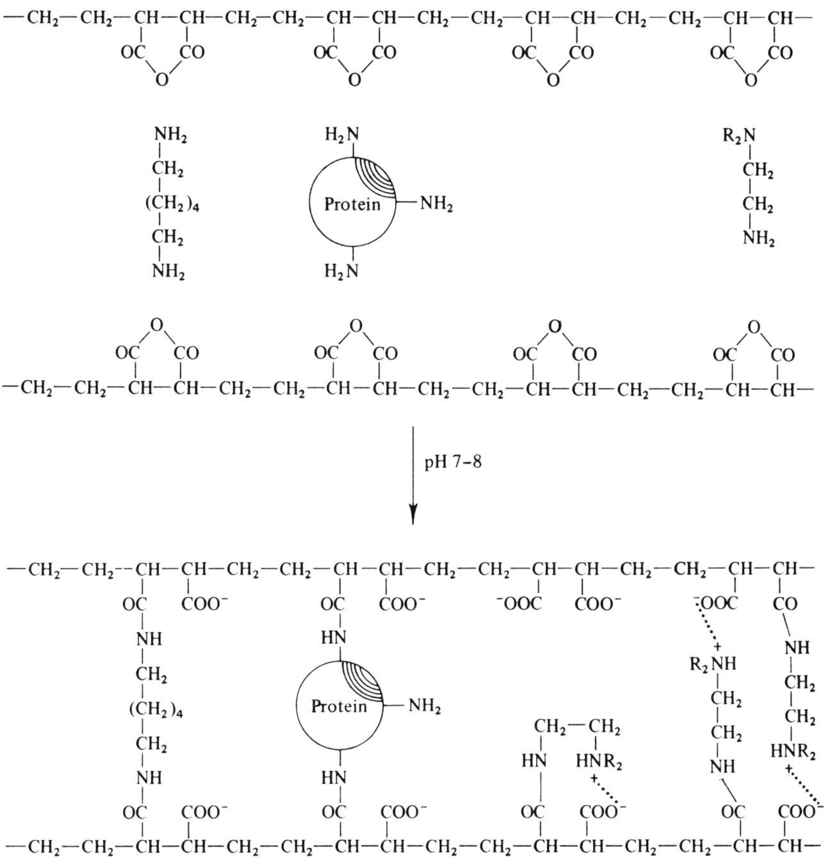

Fig. 5. Coupling of proteins to the partially neutralized resin derived from an ethylene-maleic anhydride copolymer (EMA). [Fritz et al. (1968, 1969).]

1969, 1970; Westman, 1969). EMA-Papain and EMA-chymotrypsin conjugates have been converted into polycationic or polyalcohol derivatives by coupling the polycarboxylic enzyme derivative with N,N-dimethylaminopropylamine or propanolamine, respectively, in the presence of carbodiimide (Goldstein, 1971). The polyanionic character of EMA-trypsin and EMA-chymotrypsin conjugates could also be partly neutralized by introducing, into the coupling mixture, varying amounts of N,N-dimethylaminoethylamine (Fritz et al., 1968, 1969) (Fig. 5).

Covalent binding of enzymes via their carboxyl groups, to aminoethyl cellulose or other resins containing primary aliphatic amines, could be effected by coupling the protein with the carrier in the presence of carbodiimide (Goldstein, 1971).

Enzymes possessing tyrosine residues not essential for catalytic activity might be linked to different carriers via azo bonds. Polydiazonium salts of the following carriers have been used for this purpose: p-aminobenzyl cellulose (Campbell et al., 1951; Sehon, 1967), m-aminobenzyloxymethyl ether of cellulose (Gurvich, 1957; Weliky and Weetall, 1965; Surinov and Manoylov, 1966; Davis et al., 1969), poly-p-aminostyrene (Grubhofer and Schleith, 1954; Sehon, 1967), a copolymer of p-aminophenylalanine and leucine (Bar-Eli and Katchalski, 1963; Silman et al., 1966) (Fig. 6), and S-MDA (Goldstein, 1970; Goldstein

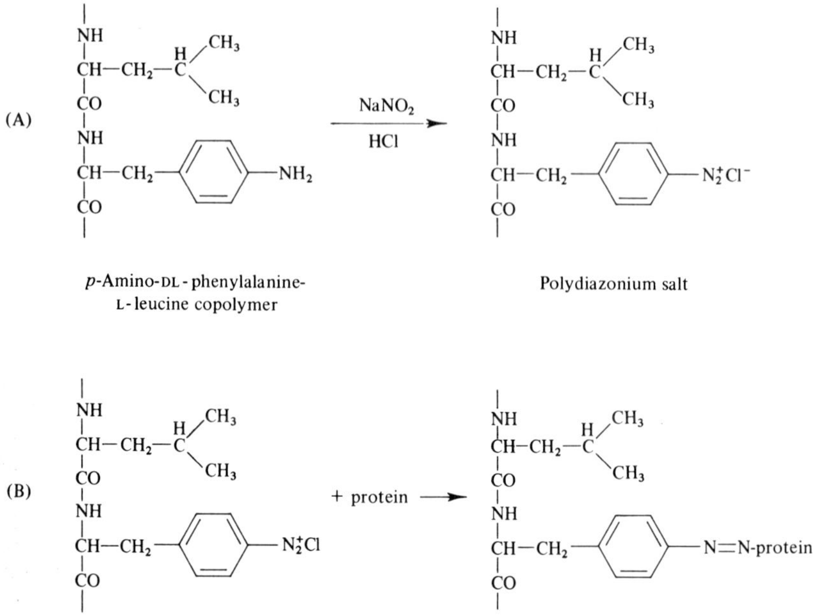

Fig. 6. Preparation of the polydiazonium salt of a p-aminophenylalanine-leucine copolymer (A), and its coupling to a protein (B). [Bar-Eli and Katchalski (1963).]

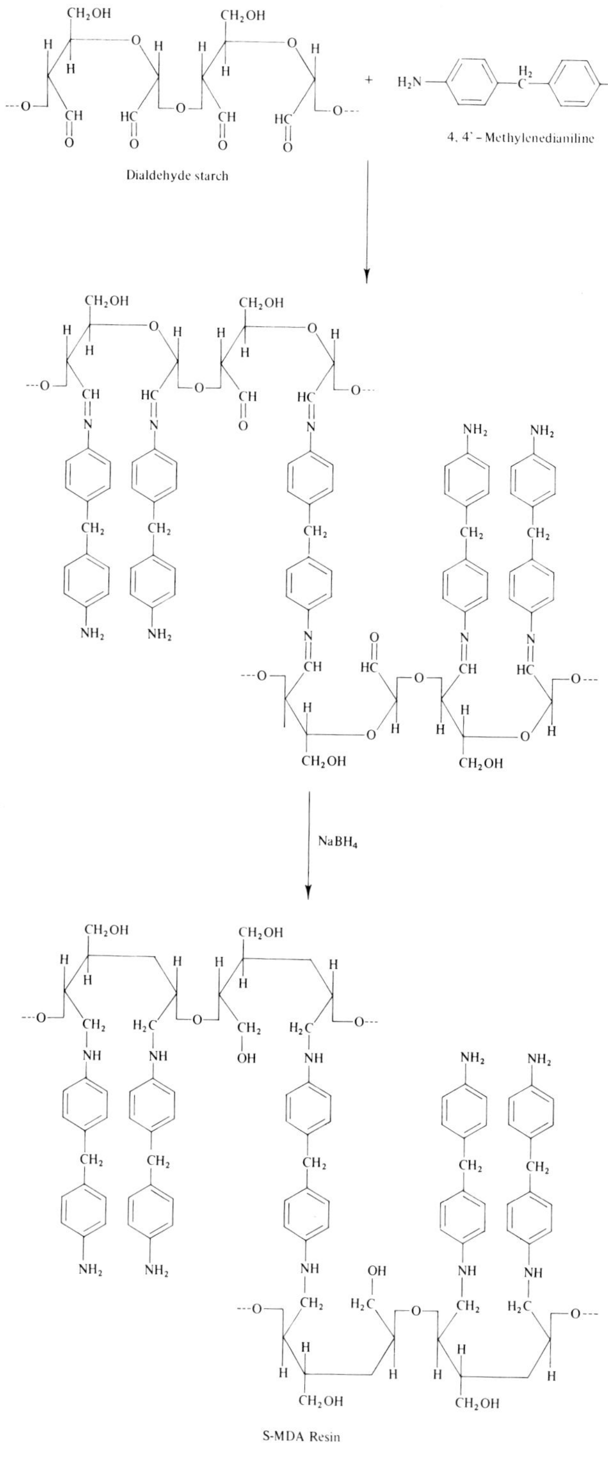

Fig. 7. Synthesis of an S-MDA resin. The resin is prepared by the condensation of dialdehyde starch with p,p'-diaminodiphenyl methane (4,4'-methylenedianiline, MDA) and the subsequent reduction of the Schiff's base polymeric product. [Goldstein et al., (1970).]

et al., 1970), a resin prepared by the condensation of dialdehyde starch (a commerically available periodate-oxidation product of starch) with p,p'-diaminodiphenyl methane (4,4'-methylenedianiline, MDA), and the subsequent reduction of the Schiff's base polymeric product (Fig. 7). Amino acid analysis of acid hydrolyzates of S-MDA derivatives of papain, subtilopeptidase A, and polytyrosyl trypsin revealed that in addition to tyrosine residues, lysine and arginine residues are attacked by the polydiazonium carrier employed (Goldstein, 1970; Goldstein et al., 1970).

A method for the chemical modification of preformed cross-linked polyacrylamide beads has been recently described by Inman and Dintzis (1969). The inert amide groups were converted into hydrazide or aminoethyl groups by reacting the polyacrylamide beads with hydrazine or with ethylene diamine (Figs. 8 and 9). The hydrazide derivative was converted into the acyl azide and coupled to proteins via amide bonds (Fig. 8). The aminoethyl derivative of polyacrylamide was converted into the p-aminobenzamidoethyl derivative which following diazotization was coupled to proteins via azo bonds (Fig. 9).

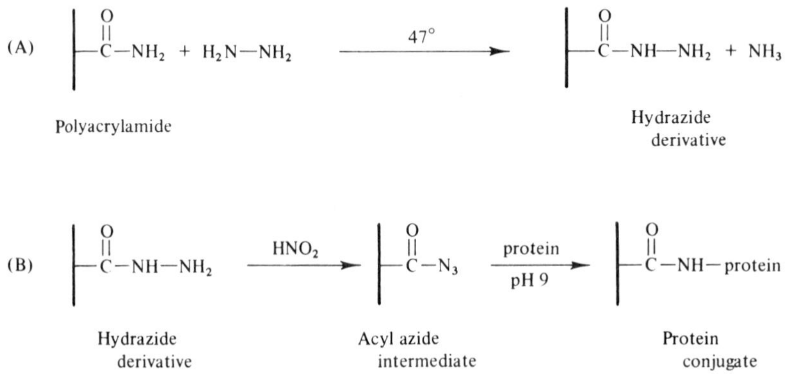

Fig. 8. Coupling of proteins to derivatized cross-linked polyacrylamide beads. (A) Reaction employed in the preparation of the hydrazide derivative of cross-linked polyacrylamide, and (B) its coupling, via the acyl azide derivative, to proteins. [Inman and Dintzis (1969).]

Most of the materials used as carriers for the covalent binding of enzymes are natural or synthetic organic polymers. These might be divided into electrically neutral carriers, such as cellulose, Sephadex, and Sepharose, and polyelectrolyte carriers such as CM-cellulose, the ethylene-maleic acid copolymers (EMA), and aminoethyl cellulose. The use of inorganic carriers such as glass has recently been reported (Weetall, 1969a). The glass was activated by coupling with γ-aminopropyl triethoxysilane, and the aminoalkylsilane-glass derivative was converted to the isothiocyanate derivative, or was reacted with *p*-nitrobenzoic acid and the nitro group reduced and diazotized.

D. Enzymes Immobilized by Intermolecular Cross-Linking

Insolubilization of enzymes and other proteins was also attained by intermolecular cross-linking making use of bifunctional reagents (Silman and Katchalski, 1966; Goldstein and Katchalski, 1968; Goldstein, 1969). Two types of cross-linking reagents have been commonly used: (a) bifunctional reagents possessing two identical functional groups such as glutaraldehyde (Richards, 1963; Quiocho and Richards, 1964; Habeeb, 1967; Ogata *et al.*, 1968; Haynes and Walsh, 1969; Jansen and Olson, 1969; Brown *et al.*, 1966; Goldman and Katchalski, 1969), bisdiazobenzidine-2,2'-disulfonic acid (Goldman *et al.*, 1965; Silman and Katchalski, 1966; Silman *et al.*, 1966); 1,5-difluoro-2, 4-dinitrobenzene (Zahn and Meienhofer, 1958; Zahn *et al.*, 1962); diphenyl-4, 4'-dithiocyanate-2,2'-disulfonic acid (Manecke and Günzel, 1967); 4,4'-

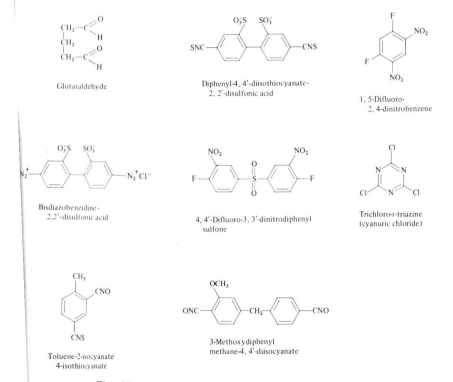

Fig. 10. Some common cross-linking reagents.

1. Water-Insoluble Enzyme Derivative

(A) $\left[\!\!\begin{array}{c}\\ \mathrm{O=C-NH_2}\\ \end{array}\!\!\right] + \mathrm{H_2N-CH_2-CH_2-NH_2} \xrightarrow{90°} \left[\!\!\begin{array}{c}\\ \mathrm{O=C-NH-CH_2-CH_2-NH_2}\\ \end{array}\!\!\right] + \mathrm{NH_3}$

Polyacrylamide   Aminoethyl derivative

(B) Aminoethyl derivative

$\xrightarrow[\mathrm{H_2O}]{\substack{1.\ \mathrm{O_2N-}\!\!\bigcirc\!\!\mathrm{-C(=O)-N_3}\\ \mathrm{DMF(Et_3N)}\\ 2.\ \mathrm{Na_2S_2O_4},\ \mathrm{H_2O}}}$ → p-Aminobenzamidoethyl derivative

3. $\mathrm{HNO_2}$ → Polydiazonium intermediate

4. protein, pH ~ 9 → $\mathrm{-NH-C(=O)-}\!\!\bigcirc\!\!\mathrm{-N=N-protein}$

difluoro-3,3′-dinitrodiphenyl sulfone (Wold, 1961); phenol-2,4-disulfonyl chloride (Herzig et al., 1964), etc., and (b) bifunctional reagents possessing two different functional groups, or groups of differing reactivities such as toluene-2-isocyanate-4-isothiocyanate (Schick and Singer, 1961); 3-methoxydiphenyl methane-4,4′-diisocyanate (Schick and Singer, 1961); trichloro-s-triazine (Surinov and Manoylov, 1966; Kay and Crook, 1967), etc. (see Fig. 10).

Immobilization of enzymes by covalent binding to a suitable support obviously eliminates leakage of activity into the medium and enables control of the physical properties and particle size of the final product. It should be pointed out, however, that the sensitivity of many enzymes to chemical modification limits the applicability of the method.

III. STABILITY OF WATER-INSOLUBLE ENZYME DERIVATIVES

The use of immobilized enzymes is primarily determined by their ability to retain enzymic activity for considerable periods of time under suitable conditions of storage. Moreover, adequate stability is essential for the repeated use of an immobilized enzyme derivative, or for its continuous utilization in column form. The nature of the carrier might affect the stability of the covalently bound protein. In principle, carriers might be found which would either increase or decrease the stability of a given protein. A carrier containing hydrophobic groups, for example, might denature the protein, analogously to a hydrophobic solvent. The proximity of a hydrophobic carrier need not effect immediate denaturation; it might cause, however, slow inactivation on storage, or modified sensitivity to heating, pH, or denaturing agents. A hydrophilic carrier of positive or negative charge might lead, under certain conditions, to an increase or a decrease in the stability of an immobilized enzyme derivative due to electrostatic interactions between bound protein and carrier. In the case of proteolytic enzymes, autodigestion often leads to inactivation. In such a system immobilization should increase stability, as the covalent binding of the enzyme molecules to the carrier should prevent their interaction with each other (Silman and Katchalski, 1966; Goldstein and Katchalski, 1968; Goldstein, 1969, 1970).

A. Storage Stability

Aqueous suspensions of ethylene-maleic acid (EMA) enzyme derivatives [e.g., trypsin (Levin et al., 1964), chymotrypsin (Goldstein, 1970), papain (Goldstein, 1970), subtilisin Carlsberg (Goldstein, 1970), subtilisin Novo (Goldstein, 1971)], CM-cellulose derivatives [e.g., chymotrypsin (Kay et al.,

1968), ficin (Hornby et al., 1966), and bromelain (Wharton et al., 1968a)] and Sepharose derivatives of chymotrypsin (Porath et al., 1967; Axén et al., 1967, 1971) could be stored at 4° for several months without significant loss of activity. Lyophilized powders of these materials retained their activity after prolonged storage at 4° and at room temperature (Levin et al., 1964; Silman and Katchalski, 1966; Goldstein and Katchalski, 1968; Goldstein, 1969, 1970).

Suspensions of insoluble enzyme derivatives utilizing p-aminobenzyl celluloses, leucine-p-aminophenylalanine copolymers and S-DMA resins (q.v.) as carriers [e.g., papain (Silman et al., 1966; Goldstein et al., 1970), polytyrosyl trypsin (Bar-Eli and Katchalski, 1963; Goldstein, 1970; Goldstein et al., 1970), urease (Riesel and Katchalski, 1964), subtilisin Carlsberg (Goldstein, 1970; Goldstein et al., 1970), and subtilisin Novo (Goldstein, 1971)] could also be stored in the cold for several months. These derivatives, however, were almost completely inactivated on lyophilization or air drying, probably due to the hydrophobic nature of the carrier (Silman and Katchalski, 1966; Goldstein, 1970; Goldstein et al., 1970).

The insoluble conjugates of invertase, pepsin, and alcohol dehydrogenase with a copolymer of methacrylic acid and methacrylic acid fluorodinitroanilide lost most of the activity within several weeks on storage at 4° (Manecke, 1962).

Enzymes occluded in polyacrylamide or starch gels (e.g. glucose oxidase, lactic dehydrogenase, acetylcholinesterase), as well as papain and alkaline phosphatase-collodion membranes, retained their activity for several months in the cold (Bernfeld and Wan, 1963; Aldrich et al., 1963, 1965; Guilbault and Kramer, 1965; Bauman et al., 1965, 1967; Goldman et al., 1965, 1968a; Wieland et al., 1966; van Duijn et al., 1967; Bernfeld et al., 1968; Pennington et al., 1968; Goldman and Katchalski, 1971). Lyophilized polyacrylamide and starch gel preparations containing the above enzymes could be easily rehydrated with the concomitant recovery of most of the enzymic activity; the enzyme-collodion membranes, however, shrank irreversibly on lyophilization and lost their activity presumably due to decreased permeability to substrate (Goldman and Katchalski, 1971). (See also Section VII.)

B. Thermal Stability

The available data on the thermal stability of immobilized enzyme systems is still rather scant. Improved thermal stabilities, as compared to the corresponding native enzymes, have been reported for CM-cellulose-ficin (Hornby et al., 1966; Lilly et al., 1966), DEAE-cellulose-lactic dehydrogenase (Wilson et al., 1968a,b), glucose oxidase immobilized on cellophane sheets (Broun et al., 1969), silastic-entrapped acetylcholinesterase (Pennington et al., 1968), and for papain attached to glass (Weetall, 1969a). These findings seem to be the exception rather than the rule. In most of the cases investigated the thermal stability of an enzyme was found to decrease upon immobilization irrespective of the type of carrier used. To give a few examples, the thermal stabilities of alkaline phosphatase (Goldman and Katchalski, 1971) and glucose-6-phosphate dehydrogenase (Goldman and Lenhoff, 1969) were drastically lowered on immobilization of these enzymes onto collodion membranes. The thermal stability of papain bound to neutral carriers such as leucine-p-aminophenylalanine copolymers (Silman et al., 1966), S-MDA resins (Goldstein, 1970; Goldstein et al., 1970), p-aminobenzyl cellulose (Goldstein, 1970; Goldstein et al., 1970), and collodion (Goldman et al., 1965, 1968a) was also lower than that of the native enzyme. A similar lowering of thermal stability has been reported for several of the polyanionic, polycationic, and polyalcoholic derivatives of papain (Goldstein, 1971).

Generally, it seems that adsorption or covalent binding of enzymes leads to a decrease in their thermal stability. This phenomenon could be possibly related to a decrease in the probability of the recovery of the native enzyme conformation following thermal perturbation.

C. Dependence of Stability on pH

Improved stabilities toward alkaline pH's (up to 10.7) have been reported for the polyanionic derivatives of several enzymes [e.g. EMA-derivatives of trypsin (Levin et al., 1964; Goldstein, 1970), chymotrypsin (Goldstein, 1970), and papain (Goldstein, 1970)]; conversely polycationic derivatives exhibited improved stability in the acid pH range (Goldstein, 1971). These phenomena could be related to local (microenvironmental, q.v.) pH effects induced by the polyelectrolyte carrier (Levin et al., 1964; Goldstein et al., 1964). (c.f. Section IV.)

The pH dependence of the stability of papain bound to the neutral carriers derived from p-aminobenzyl cellulose (Goldstein, 1970; Goldstein et al., 1970), leucine-p-aminophenylalanine copolymers (Silman et al., 1966), and S-MDA resins (Goldstein, 1970; Goldstein et al., 1970) (q.v.) was similar to that of the native enzyme. Derivatives of subtilisin Carlsberg bound to the same carriers showed increased stability in the acid pH region (pH 3–4) (Goldstein, 1970; Goldstein et al., 1970). p-Aminobenzyl cellulose and S-MDA derivatives of polytyrosyl trypsin (Goldstein et al., 1970) exhibited maximal stability at neutral and alkaline pH values, in contradistinction to native trypsin or polytyrosyl trypsin which are most stable at low pH values (Manecke, 1964; Goldstein and Katchalski, 1968; Goldstein, 1969).

In conclusion it should be indicated that the modified stability pattern of covalently bound enzymes is determined not only by the physical and chemical characteristics of the carrier, but also by the nature of the chemical modification of the enzyme moiety brought about by the covalent binding.

IV. KINETIC BEHAVIOR OF PARTICULATE IMMOBILIZED ENZYME SYSTEMS

The kinetic behavior of immobilized enzyme systems is the result of the superposition of the characteristics of the carrier upon the specific kinetic parameters of the enzyme. The effects of immobilization on the apparent kinetic behavior might be resolved as follows. (a) Effects of diffusion limitations—the kinetic parameters of enzymic reactions occurring in the heterogeneous phase, particularly where enzyme columns or enzyme membranes are concerned might be determined to a considerable extent by the rate of diffusion of substrate across the unstirred layer surrounding the insoluble matrix. (b) Steric effects—when an immobilized enzyme acts on a high molecular weight substrate, steric restrictions imposed by the matrix might markedly affect the course of the catalytic reaction. (c) Effects of enzyme modification—these result from the chemical modification of the enzyme caused by the process of covalent binding. Chemical modifications lead in many cases to an alteration of the overall net charge of the enzyme, as well as to specific neighbor effects on the catalytic site of the enzyme. (d) Microenvironmental effects—these are most pronounced in the case of polyelectrolyte carriers, where the charged matrix imposes a modified microenvironment on the immobilized enzyme. Other microenvironmental effects such as the effect of carrier on the dielectric constant of the immobilized enzyme phase, or its effect on the local solubility of substrate and product, should also be considered.

A. Effects of Diffusion Limitations

The covalent attachment of enzymes to a great variety of electrically neutral matrices has been described in the literature. Organic and inorganic carriers, such as cellulose, Sephadex, Sepharose, polyacrylamide, diazotized synthetic resins, silica gel, and glass (see Section II), have been commonly used. The physical texture of these materials varies from rigid, dense, completely insoluble particles, to soft, highly swollen cross-linked gels. Rigid particles can bind protein only to their outer surface, whereas swollen gel particles can bind protein within their entire volume. The binding capacity of swollen gel particles is thus a function of the total number of reactive groups; the binding capacity of rigid particles, on the other hand, is mainly determined by their overall surface area.

In nonenzymic heterogeneous catalysis the rate of diffusion of the reactants toward the active surface of the catalyst has been shown to play a significant role in determining the kinetics of the reaction (Thiele, 1939; Wheeler, 1951; Helfferich, 1962b). In analogy the rate of diffusion of the substrate has been found to affect the apparent kinetic parameters of immobilized enzyme systems.

Immobilized enzyme particles in aqueous suspension are surrounded by an unstirred layer of solvent, the thickness of which is determined by the rate of stirring (Nernst, 1904; Helfferich, 1962b). In the course of an enzymic reaction a concentration gradient of substrate is established across the unstirred layer (Goldman and Katchalski, 1971). Saturation of the immobilized enzyme will thus occur at substrate concentrations which are higher than those required for the saturation of the corresponding native enzyme in solution. This will lead to an increase in the value of the apparent Michaelis constant, K_m'. In swollen insoluble enzyme particles in which the enzyme is distributed throughout the particle, concentration gradients of substrate will also be established within the domain of the immobilized enzyme phase. Full activity will be attained only when the local substrate concentration markedly exceeds the K_m value of the native enzyme (Goldman et al., 1968b). In this respect a swollen enzyme particle resembles enzyme membranes, the kinetics of which are described in detail in Section VI.

The effect of an unstirred layer surrounding a rigid particle on the Michaelis constant of an attached enzyme is illustrated in the following examples: Hornby et al. (1968) reported a K_m' of 5.6×10^{-4} M for chymotrypsin attached to CM-cellulose using acetyl-L-tyrosine ethyl ester as substrate. A value of $K_m = 2.7 \times 10^{-4}$ M was given by the same authors for native chymotrypsin. Goldstein et al. (1970; Goldstein, 1970) found, for S-MDA-papain conjugates acting on benzoylglycine ethyl ester, a K_m' of 3.4×10^{-2} M. The K_m for papain was 1.8×10^{-2} M. A $K_m' = 1.7 \times 10^{-2}$ M was found for S-MDA-subtilopeptidase A conjugates acting on acetyl-L-tyrosine ethyl ester (Goldstein, 1970; Goldstein et al., 1970). The K_m recorded for native subtilopeptidase A was 0.54×10^{-2} M.

The effects of diffusion of substrate into a swollen matrix on the apparent kinetic parameters of an immobilized enzyme have been demonstrated by Axén et al. (1970), who studied the kinetics of chymotrypsin-Sepharose conjugates. The values of the apparent Michaelis constant of the immoblized chymotrypsin preparations acting on acetyl-L-tyrosine ethyl ester were about tenfold higher than that obtained for the native enzyme. On solubilization of the chymotrypsin-Sepharose conjugate, by digestion of the carrier with dextranase, the values of K_m' dropped to essentially the K_m of native chymotrypsin.

B. Steric Effects

Immobilization of an enzyme may lead to steric restrictions on its availability to high molecular weight substrates. The data on the kinetic behavior of immobilized enzymes toward high molecular weight substrates is essentially limited to the proteases. The specific activity of immobilized proteases toward proteins, calculated from the initial rates of hydrolysis, is usually lower than that of the corresponding native enzymes, the amount of bound enzyme being determined by a rate assay using low molecular weight substrate (Silman and Katchalski, 1966; Goldstein and Katchalski, 1968; Goldstein, 1969, 1970; Goldstein et al., 1970). Water-insoluble derivatives of polytyrosyl trypsin (Bar-Eli and Katchalski, 1963; Goldstein et al., 1970) and papain (Silman et al., 1966; Goldstein et al., 1970), for example, hydrolyzed casein at initial rates which were 30% and 50% of those expected on the basis of their activity toward benzoyl-L-arginine ethyl ester. A low proteolytic activity was also reported for the CM-cellulose conjugates of ficin (Hornby et al., 1966; Lilly et al., 1966), bromelain (Wharton et al., 1968a), and trypsin (Mitz and Summaria, 1961), and for polyanionic, polycationic, polyalcohol, and S-MDA derivatives of papain (Goldstein, 1970; Goldstein et al., 1970) and subtilopeptidase A (Goldstein, 1970; Goldstein et al, 1970). In the majority of cases the lowering in proteolytic activity could be attributed to steric hindrance induced by the carrier (Silman and Katchalski, 1966; Goldstein and Katchalski, 1968; Goldstein, 1969, 1970; Goldstein et al., 1970). The lowering of the initial rate of hydrolysis of proteins by immobilized proteases is often accompanied by a decrease in the total number of peptide bonds susceptible to hydrolysis (Levin et al., 1964; Silman and Katchalski, 1966; Alexander et al., 1966; Ong et al., 1966; Lowey et al., 1966, 1967, 1968; Slayter and Lowey, 1967; Lowey, 1968; Goldstein and Katchalski, 1968; Goldstein, 1969, 1970; Westman, 1969; Goldstein et al., 1970). This effect is particularly pronounced in the case of polyelectrolyte enzyme derivatives where electrostatic interactions between the charged carrier and the protein substrate are superimposed on the steric restrictions (Levin et al., 1964; Alexander et al., 1965; Silman and Katchalski, 1966; Ong et al., 1966; Lowey et al., 1966, 1967, 1968; Lowey, 1968; Slayter and Lowey, 1967; Goldstein and Katchalski, 1968; Goldstein, 1969). A detailed description of these effects is given in Section VII.

C. Effects of Chemical Modification

The covalent binding of an enzyme to a polymeric carrier might affect the kinetic behavior of the bound enzyme as a result of alteration in its net charge, nearest neighbor effects on the active site region, and perturbations in intramolecular interactions. These effects can be simulated by analogous chemical modifications with low molecular weight reagents. They can be readily detected in systems in which electrically neutral carriers are employed. In the case of polyelectrolyte carriers the effects of chemical modification are usually masked by the considerably larger electrostatic effects induced by the charged carrier (Silman and Katchalski, 1966; Goldstein and Katchalski, 1968; Wharton et al., 1968a,b; Goldstein, 1969).

The amount of data available on the effects of chemical modification of proteolytic enzymes by reagents of low molecular weight on their kinetic parameters is rather limited (Sri Ram et al., 1954, 1962; Vallee and Riordan, 1969). Acetylation, succinylation, or reaction with ethyl iminocarbonate of chymotrypsin (Sri Ram et al., 1954; Axén et al., 1970; Goldstein, 1971) results in a displacement of the pH activity curve of the modified enzyme toward more alkaline pH values as compared to native chymotrypsin. The shifts observed at low ionic strength (0.6 to 1.0 pH units) could be abolished at high ionic strengths. This phenomenon is most likely due to the increase in the net negative charge of the protein resulting from the blocking of amino groups and a corresponding increase in the pK_a of the histidine imidazole at the active site.

Displaced pH activity profiles have been reported for papain (Goldstein et al., 1970), subtilopeptidase A (Goldstein et al., 1970), and polytyrosyl-trypsin (Goldstein et al., 1970) linked via azo bonds to neutral carriers, such as S-MDA resins (Goldstein, 1970; Goldstein et al., 1970), p-aminobenzyl cellulose (Goldstein et al., 1970), and leucine-p-aminophenylalanine copolymers (Bar Eli and Katchalski, 1963). The pH-activity curves recorded were essentially independent of ionic strength. In view of the fact that polydiazonium carriers react with tyrosine, lysine, and arginine residues, an increase in the net negative charge of the bound enzymes might be expected. However the ionic strength independence of the displaced pH activity curves recorded, suggests specific, short range interactions within the region of the active site of the enzymes.

D. Microenvironmental Effects

The most thoroughly investigated effects of microenvironment on the kinetic behavior of immobilized enzymes are those of an electrostatic field produced by highly charged carriers.

Goldstein et al. (1964; Goldstein, 1970) have shown that the pH activity profiles of the polyanionic derivatives of several proteolytic enzymes acting on their specific low molecular weight substrates are displaced toward more alkaline pH values by 1–2.5 pH units, at low ionic strength ($\Gamma/2 \approx 0.01$) as compared to the native enzymes. Polycationic derivatives of the same enzymes exhibit the reverse effect, i.e., displacement of the pH activity profile toward more acidic pH values (Goldstein and Katchalski, 1968; Goldstein, 1971). These anomalies are abolished at high ionic strength ($\Gamma/2 \geqslant 1$). To illustrate this phenomenon, the pH activity profiles of a polyanionic derivative of trypsin (EMA-trypsin) and of the polyanionic and polycationic derivatives of chymotrypsin [EMA-chymotrypsin and polyornithyl chymotrypsin (Goldstein and Katchalski, 1968)] are shown in Figs. 11 and 12. Charged derivatives of papain (Goldstein, 1970), ficin (Hornby et al., 1966), and subtilopeptidase A (subtilisin Carlsberg) (Goldstein, 1970) exhibit similar behavior. Furthermore, the apparent Michaelis

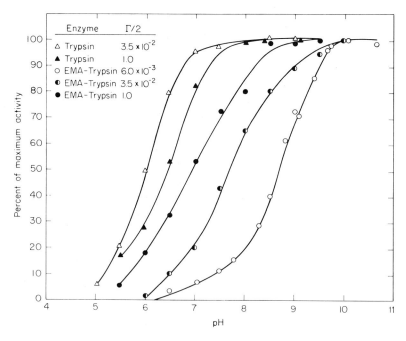

Fig. 11. pH Activity curves for trypsin and a polyanionic, ethylene-maleic acid copolymer derivative of trypsin (EMA-trypsin) at different ionic strengths, using benzoyl-L-arginine ethyl ester as substrate [redrawn from the data of Goldstein et al. (1964)].

constant of a polyanionic derivative of trypsin (EMA-trypsin), using the positively charged substrate benzoyl-L-arginine amide (BAA) ($K_m' = 2 \times 10^{-4}$ M), was lower by more than one order of magnitude at low ionic strength, as compared with that of the native enzyme (Goldstein et al., 1964) ($K_m = 6.9 \times 10^{-3} M$) (Figs. 13 and 14). Similar effects have been reported for the polyanionic derivatives of papain (EMA-papain) (Goldstein, 1970), ficin (Hornby et al., 1966) (CM-cellulose-ficin), and bromelain (CM-cellulose-bromelain) (Wharton et al., 1968a,b) using benzoyl-L-arginine ethyl ester as substrate. The perturbation of the apparent Michaelis constant is abolished at high ionic strength (Goldstein et al., 1964; Wharton et al., 1968b) (see Fig. 14). K_m values similar to those of the native enzyme have been reported for the polyanionic and polycationic derivatives of chymotrypsin and for a polyanionic derivative of papain (Goldstein and Katchalski, 1968; Goldstein, 1969, 1970) using uncharged substrates. Acetyl-L-tyrosine ethyl ester and benzoylglycine ethyl ester were the substrates used with the chymotrypsin and papain derivatives, respectively.

Goldstein et al. (1964) showed that the above phenomena could result from an unequal distribution of hydrogen and hydroxyl ions and of charged substrates

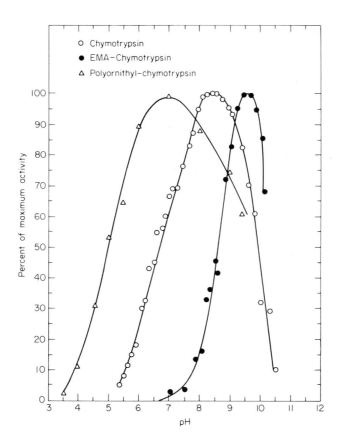

Fig. 12. pH Activity curves at low ionic strength ($\Gamma/2 = 0.008$) for chymotrypsin, a polyanionic derivative of chymotrypsin (EMA-chymotrypsin) and a polycationic, polyornithyl derivative of chymotrypsin, using acetyl-L-tyrosine ethyl ester as substrate. [Goldstein and Katchalski (1968).]

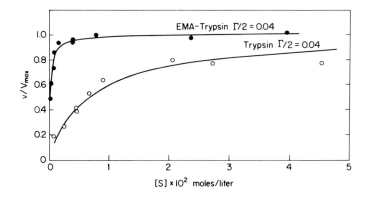

Fig. 13. Normalized Michaelis-Menten plots for trypsin and a polyanionic derivative of trypsin (EMA-trypsin) acting on benzoyl-L-arginine amide. [Goldstein *et al.* (1964).]

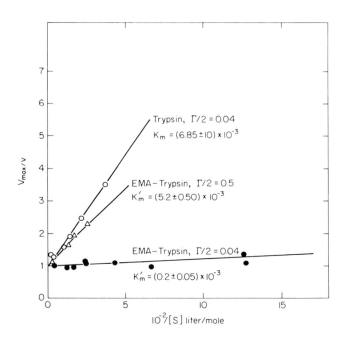

Fig. 14. Normalized Lineweaver-Burk plots for trypsin and a polyanionic derivative of trypsin (EMA-trypsin) acting on benzoyl-L-arginine amide. [Goldstein *et al.* (1964).]

between the "polyelectrolyte phase," within which the immobilized enzyme is embedded, and the outer solution.

The local hydrogen ion concentration in the domain of a charged enzyme derivative (see Fig. 15) can be described, assuming a Maxwell-Boltzmann distribution, by Eq. (1).

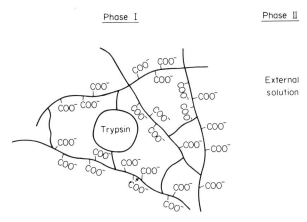

Fig. 15. Schematic presentation of a polyanionic enzyme conjugate in suspension. [Goldstein *et al.* (1964).]

$$a_{H^+}^i = a_{H^+}^o \exp(z\varepsilon\psi/kT) \qquad (1)$$

where $a_{H^+}^i$ and $a_{H^+}^o$ are the hydrogen ion activities in the polyelectrolyte-enzyme derivative phase (phase I) and the outer solution (phase II), ψ is the electrostatic potential in the domain of the charged immobilized enzyme particle, ε is the positive electron charge, z is a positive or negative integer of value unity in the case of hydrogen ions, k is the Boltzman constant, and T is the absolute temperature. Equation (1) shows that the local pH in the domain of a polyanionic enzyme derivative is lower than that measured in the external solution. The reverse is true for a polycationic enzyme derivative. Consequently, the pH activity profile of an enzyme immobilized within a charged carrier is displaced toward more alkaline or toward more acid pH values for a negatively or positively charged carrier, respectively. Quantitatively this might be expressed in the form:

$$\Delta pH = pH^i - pH^o = 0.43z\varepsilon\psi/kT \qquad (2)$$

where ΔpH is the difference between the local pH within the polyelectrolyte-enzyme phase (pH^i) and the pH of the outer solution (pH^o).

The pH of the outer solution can be measured potentiometrically with a standard glass electrode. The pH of the polyelectrolyte phase, pH^i, cannot be measured directly; it can, however, be obtained from a comparison of the normalized pH activity profiles of the polyelectrolyte-bound enzyme and the corresponding free enzyme (e.g. Figs. 11 and 12); pH^i is equal to the pH at which the native enzyme shows a catalytic activity identical with that of the bound enzyme under the conditions specified. The experimentally determined values of pH^i and pH^o allow, by means of Eq. (2), the calculation of the electrostatic potential ψ, prevailing in the polyelectrolyte-enzyme phase. The values of ψ (in the range of 50–150 mV) calculated for ethylene-maleic acid copolymer derivatives of trypsin and chymotrypsin (EMA-trypsin and EMA-chymotrypsin) and for polyornithyl derivatives of chymotrypsin (Figs. 11 and 12) were in good agreement with the ψ values expected from polyelectrolyte theory for the corresponding ionized polymers.

The dependence of enzymic activity on pH is commonly ascribed to the dissociation of ionizing groups participating in the enzymic catalysis (Gutfreund, 1955; Bender and Kézdy, 1965). The chemical identity of such an active site ionizing group has, in many cases, been deduced from the value of the dissociation constant, pK_a(app), calculated from the pH activity profile of the enzyme (Dixon and Webb, 1964). Thus, the dissociation constants calculated from the acid limbs of the pH rate curves of trypsin, subtilopeptidase A, and chymotrypsin [pK_a(app) $\approx$ 7] have been assigned to active site histidines (Bender and Kézdy, 1965). In the case of chymotrypsin, the hypothetical histidine residue has been unequivocally identified as histidine-57 on the basis of both chemical and crystallographic evidence (Dixon and Webb, 1964; Siegler et al., 1968). The displaced pH activity profiles of a polyelectrolyte enzyme derivative can therefore be alternatively represented in terms of changes in the values of the apparent acidic dissociation constants [pK_a(app)] of the "active site" ionizing group effected by the polyelectrolyte "microenvironment" of the enzyme derivative.

The changes in the values of the apparent Michaelis constants (K_m') of polyelectrolyte enzyme derivatives acting on charged low molecular weight substrates (c.f. Figs. 13 and 14) could be related to the unequal distribution [c.f. Eq. (1)] of substrate between the charged enzyme particle and the outer solution [Eq. (3)] (Goldstein et al., 1964). The relation between the concentration of substrate in the domain of the enzyme-polyelectrolyte conjugate, S^i, and the substrate concentration in the outer solution, S^o, is given, assuming a Maxwell-Boltzmann distribution, by Eq. (3):

$$S^i = S^o \exp\left(z\varepsilon\psi/kT\right) \qquad (3)$$

Equation (3) shows that $S^i > S^o$ when the polyelectrolyte-enzyme conjugate and the substrate are of opposite charge. The enzyme derivative will thus attain the limiting rate, $V_{\max}$, at a lower bulk concentration of substrate, S^o, as compared to the native enzyme; the apparent Michaelis constant for the immobilized enzyme (K_m') will therefore be lower than that of the corresponding native enzyme

(K_m). For substrate and polyelectrolyte enzyme conjugate of the same charge the opposite will be true, i.e. $S^i < S^o$ and the value of K'_m of the enzyme derivative will be higher than K_m.

The quantitative relation between the apparent Michaelis constant of a charged enzyme derivative, K'_m, the Michaelis constant of the native enzyme, K_m, and the electrostatic potential ψ, can be derived as follows (Goldstein et al., 1964): insertion of Eq. (3) into the Michaelis-Menten equation [(Eq. (4)] gives Eq. (5).

$$v = -\frac{dS}{dt} = \frac{V_{\max} S}{K_m + S} \tag{4}$$

$$v' = \frac{V_{\max} S^o \exp(z\varepsilon\psi/kT)}{K_m + S^o \exp(z\varepsilon\psi/kT)} \tag{5}$$

where v' denotes the velocity of the enzyme-polyelectrolyte derivative catalyzed reaction.

It follows from Eq. (5) that $v' = \frac{1}{2}V_{\max}$ when $S^o = K_m \exp(-z\varepsilon\psi/kT)$. Thus the value of the outer substrate concentration, S^o, at which half-maximum velocity is attained, leads to an apparent Michaelis constant, K'_m, related to the Michaelis constant of the native enzyme, K_m, by the expression

$$K'_m = K_m \exp(-z\varepsilon\psi/kT) \tag{6}$$

Equation (6) can be rewritten as

$$\Delta pK_m = pK'_m - pK_m = \log(K_m/K'_m) = 0.43 z\varepsilon\psi/kT \tag{7}$$

The values of ψ calculated from ΔpK_m, by means of Eq. (7), for EMA-trypsin acting on benzoyl-L-arginine amide (BAA) at low ionic strength ($\psi = 92$ mV at $\Gamma/2 = 0.04$; see data of Fig. 14), were found to be in good agreement with the ψ values calculated by Eq. (2) from the pH activity profiles of EMA-trypsin acting on benzoyl-L-arginine ethyl ester (BAEE) at the same ionic strength ($\psi = 96$ mV at $\Gamma/2 = 0.035$; see data of Fig. 11).

Using the Donnan relationship (Tanford, 1961) to describe the distribution of charged substrate between the outer solution and the polyelectrolyte enzyme phase, Wharton et al. (1968b) deduced Eq. (8) in which the effect of ionic strength, I, on the apparent Michaelis constant, K'_m, is given explicitly.

$$(K'_m)^2 = \gamma^2 K_m [K_m - K'_m Z m_c / I] \tag{8}$$

In this equation Z is the modulus of the number of charges on the matrix and m_c the concentration of the matrix in its own hydrated volume. Zm_c thus denotes the effective concentration of fixed charged groups in the polyelectrolyte phase. The activity coefficient, γ, is given by $\gamma = \gamma^i_\pm / \gamma^o_\pm$, where $\gamma^i_\pm$ and $\gamma^o_\pm$ are the mean ion activity coefficients of the matrix and the outer phase, respectively.

Rearrangement of Eq. (8) gives

$$K'_m = \gamma K_m [1 - K'_m Z m_c / K_m I]^{1/2} \tag{9}$$

Binomial expansion and truncation after the first term yields

$$K'_m = \gamma K_m [1 - K'_m Z m_c / 2 K_m I] \tag{10}$$

Equation (10) is valid only if $0 < (K'_m Z m_c / 2 K_m I) < 1$.

Since $K_m / K'_m = S^i / S^o$ [see Eqs. (3) and (6)] a comparison of Eqs. (10) and (6) shows that both equations will yield identical expressions for K_m / K'_m when

$$\exp(z\varepsilon\psi/kT) = 1/\gamma + Z m_c / 2I \tag{11}$$

Equation (6) is the more general of the two; it does not, however, permit an assessment of the perturbing effect of the charged matrix on the apparent Michaelis constant, K'_m, in terms of readily measurable quantities. Equation (10), on the other hand, relates the apparent Michaelis constant, K'_m, with the

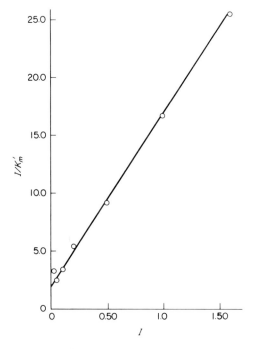

Fig. 16. Dependence of the values of the apparent Michaelis constant, K'_m, upon the ionic strength, I, for the carboxymethyl cellulose-bromelain catalyzed hydrolysis of benzoyl-L-arginine ethyl ester. Solid line calculated according to Eq. (12) $1/K'_m = I/\gamma K_m + Z m_c / 2 K_m$, using the values $\gamma K_m = 0.066 \pm 0.002$ and $\gamma Z m_c = 0.26 \pm 0.04$ [Wharton et al. (1968b).]

ionic strength of the medium, I, and the electrostatic parameter, Zm_c, characteristic of the matrix material. Zm_c, the effective concentration of fixed charged groups in the hydrated volume of the polyelectrolyte matrix, can be estimated by suspending a given amount of matrix material, of known net charge, in water and allowing it to settle in a measuring cylinder. The hydrated volume of the solid can then be read off on the cylinder.

Equation (10) can be rearranged into Eqs. (12) and (13)

$$I/K'_m = I/\gamma K_m + Zm_c/2K_m \qquad (12)$$

$$K'_m = \gamma K_m I/[(\gamma Zm_c/2) + I] \qquad (13)$$

Equation (12) predicts that a plot of I/K'_m against I will be linear, assuming that γ is constant, with intercept $Zm_c/2K_m$ and slope $1/\gamma K_m$ (see Fig. 16). Equation (13) is the hyperbolic form of Eq. (10) (see Fig. 17).

The validity of Eqs. (10), (12), and (13) was illustrated in a study of the kinetics of hydrolysis of benzoyl-L-arginine ethyl ester (BAEE) by carboxymethyl cellulose-bromelain (Wharton et al., 1968b). The value of K_m obtained for the hydrolysis of BAEE by native bromelain was invariant with ionic strength. The value of K'_m at low ionic strength ($K'_m = 0.007$ M, at $I = 0.023$) was found to be lower by about one order of magnitude than the value of K_m for native bromelain under similar conditions ($K_m = 0.11$ M). K'_m increased with increasing

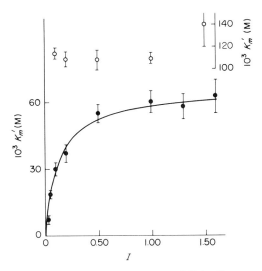

Fig. 17. Dependence of the values of the apparent Michaelis constant, K'_m, upon the ionic strength, I, for the hydrolysis of benzoyl-L-arginine ethyl ester by bromelain (○) and by carboxymethyl cellulose-bromelain (●). The solid curve (for CM-cellulose-bromelain) was calculated according to Eq. (13) $K'_m = \gamma K_m I/(\gamma Zm_c/2 + I)$, with $\gamma K_m = 0.066$ and $\gamma Zm_c = 0.26$. [Wharton et al. (1968b).]

ionic strength, approaching a value which is somewhat lower than that of the native enzyme. The experimental data plotted according to Eqs. (12) and (13) are shown in Figs. 16 and 17, respectively.

Equation (13) shows that $K_m' = \gamma K_m/2$ when $I = \gamma Z m_c/2$. Thus, if it is assumed that $\gamma = 1$, $K_m' = K_m/2$ when $I = Zm_c/2$, i.e. the apparent K_m is numerically equal to half its limiting value when the ionic strength of the outer solution is numerically equal to half the concentration of CM-cellulose carboxylate groups in their own hydrated volume.

The finding that the limiting value of K_m' (at very high ionic strength) is lower than the value of K_m for the native enzyme suggests that interactions of matrix and substrate other than those of the charge-charge type might be of some significance in determining the magnitude of the apparent Michaelis constant.

E. Immobilized Multienzyme Systems

Pathways of intermediary metabolism can be considered as sets of enzymic reactions advancing in linear or cyclic sequences. Limited experimental data are available on the kinetics and mode of action of membrane-bound enzymes performing sets of consecutive reactions. Two recent publications dealing with the preparation and kinetic behavior of matrix-bound two-enzyme systems catalyzing consecutive reactions are briefly summarized below.

The overall kinetics of an immobilized two-enzyme system consisting of hexokinase (HK) and glucose-6-phosphate dehydrogenase (G-6-PDH), covalently bound to Sepharose or acrylic acid-acrylamide beads, have been investigated by Mosbach and Mattiasson (1970). The enzymic activity of the matrix-bound two enzyme system was determined in a coupled test, in the presence of glucose and ATP, the substrate for hexokinase, and NADP$^+$ which is reduced to NADPH by the dehydrogenase (G-6-PDH). The overall rate of the consecutive set of reactions was expressed as moles of NADPH formed per minute. Subsequently the separate enzymic activities of bound hexokinase and glucose-6-phosphate dehydrogenase were determined. The kinetics of the immobilized systems was then compared with the corresponding homogeneous systems in which the native free enzymes were used. Taking the rate of formation of NADPH in the soluble system as reference, a 100–140% increase in the rate of NADPH production by the matrix-bound two-enzyme system was recorded. Moreover, it was found that the concentration of NADPH increased linearly with time in the case of the immobilized HK/G-6-PDH system, whereas a distinct time lag in the formation of NADPH was observed with the soluble system. These observations were interpreted as being due to the spacial proximity of the two enzymes on the supporting matrix, leading to higher local concentrations of the intermediate product, G-6-P, formed by the first enzyme, hexokinase, than those to be expected in the soluble system.

The kinetic behavior of a two-enzyme membrane-supported system which carries out two consecutive reactions has been recently examined by Goldman and Katchalski (1971). The analysis was based on the assumption that a quasi-stationary state was established within the unstirred layer at the two-enzyme membrane/solution interface. Under these conditions, the rate of flow of the product of the first enzymic reaction into the bulk of the solution equals the difference between the rate of its production and the rate of its consumption by the second enzyme. The behavior of the immobilized two-enzyme system was compared to the analogous soluble system consisting of the two enzymes in homogeneous solution. Using the boundary conditions: (a) the activity of the first enzyme is independent of the second, (b) the enzymic reactions follow first order kinetics, (c) the concentration of the substrate of enzyme 1 is constant throughout the enzymic reaction, (d) the concentrations of the products of both enzymic reactions are time dependent, and (e) the system is of finite volume. Goldman and Katchalski (1971) showed that for the immobilized system the concentrations of the products of enzyme 1 and 2 in the bulk of the solution, P_b^1 and P_b^2, increased linearly with time for all hypothetical systems studied. For systems which consisted of enzymes of relatively high activity, the enzymic reactions were diffusion controlled, i.e., the activity of enzyme 1 was limited by the rate of diffusion of substrate from the bulk of the solution, and the activity of enzyme 2 approached that exhibited by enzyme 1. Kinetic analysis of the corresponding homogeneous system containing both enzymes in solution revealed the existence of an initial lag period in the production of the final product P_b^2. The kinetic analysis also showed that for the immobilized enzyme system the rate of production of end product in the initial stages of the reaction was considerably higher than that calculated for an analogous soluble, homogeneous system.

These theoretical predictions are in good agreement with the data of Mosbach and Mattiasson (1970) on immobilized hexokinase + glucose-6-phosphate dehydrogenase described above.

V. ENZYME COLUMNS

Columns of immobilized enzymes can be employed for the continuous preparation of product, for regulation of the extent of conversion of substrate to product, and in automated analytical procedures (Silman and Katchalski, 1966; Goldstein and Katchalski, 1968; Goldstein, 1969).

The kinetic behavior of enzyme columns has been investigated by several authors (Bar-Eli and Katchalski, 1963; Lilly et al., 1966). Bar-Eli and Katchalski

(1963) have derived an expression [Eq. (15)], correlating the extent of substrate conversion with the enzyme concentration, E^o, the height of the column, h, and the linear rate of flow of substrate through the column, V (in centimeters per unit time). Their equation was obtained from the integrated form of the Michaelis-Menten equation [Eq. (14)], derived for enzymes in solution:

$$(S_o - S_t) + K_m \ln(S_o/S_t) = k_3 E_o t \qquad (14)$$

S_o is the initial concentration of substrate, S_t the concentration of substrate at time t, and k_3 and K_m have their usual meaning. Substitution of the time, t, in Eq. (14) by h/V, the residence time of substrate in the enzyme column and S_t by S_h, the concentration of substrate in the eluent, yields

$$(S_o - S_h) + K_m \ln(S_o/S_t) = k_3 E_o(h/V) \qquad (15)$$

S_o in Eq. (15) denotes the concentration of substrate entering the column. Equation (15) shows that the extent of conversion of substrate is determined by the effective concentration of enzyme, by the height of the column and by the linear rate of flow of substrate through the column.

When $S \gg K_m$, integration of the Michaelis-Menten equation (Eq. 4) yields

$$(S_o - S_h)/S_o = k_3 E_o h/S_o V \qquad (16)$$

Equation (16) was verified for a water-insoluble trypsin derivative packed in a column using L-arginine ethyl ester as substrate. In the presence of a relatively large excess of substrate, the amount of product was inversely proportional to the height of the column.

Equation (17), analogous to Eq. (15), was derived by Lilly *et al.* (1966) to describe the kinetic behavior of an immobilized enzyme packed in a column.

$$PS_o - K'_m \ln(1 - P) = k_3 E_t \beta/Q \qquad (17)$$

In this equation, E_t is the total amount of enzyme and Q is the flow rate, in units of volume of eluent per unit time. Using V_l to denote the void volume of the column, and V_t the total volume, the residence time of substrate in the column is $t = V_l/Q$ and the voidage of the column is $\beta = V_l/V_t$. The reaction capacity of the column is given by $C = k_3 E_t \beta$, and the fraction of substrate reacted in the column by $P = (S_o - S_t)/S_o$.

Equation (17) may be rearranged to give

$$PS_o = K'_m \ln(1 - P) + C/Q \qquad (18)$$

Equation (18) shows that if values of P are measured when various initial concentrations of substrate are perfused through an enzyme column at an identical flow rate (i.e. $Q = $ const), PS_o plotted against $\ln(1 - P)$ will give a straight line if K'_m and C are constant at this flow rate. The slope of the line will be equal to K'_m and the intercept to C/Q. Thus K'_m and C can be determined for any flow rate through the column. Kinetic data on the hydrolysis of benzoyl-L-arginine ethyl ester in packed columns of carboxymethyl cellulose-ficin when plotted according to Eq. (18) indicated that the apparent Michaelis constant (K'_m) decreased with increasing flow rate and asymptoted toward a minimal value at high flow rates (Lilly et al., 1966) (see Figs. 18 and 19). In general the value of

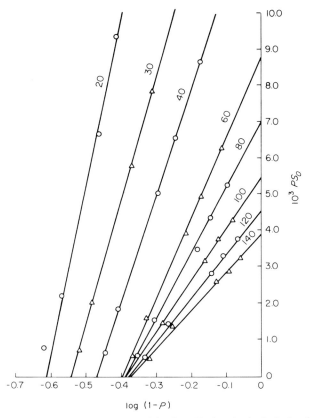

Fig. 18. Relationship between PS_0 and $\log(1 - P)$, for the hydrolysis of benzoyl-L-arginine ethyl ester by a column of carboxymethyl cellulose-ficin (see text). The numbers on the curves indicate the flow rate, Q, in milliliters per hour. [Lilly et al. (1966).]

the Michaelis constant was higher than that observed under comparable conditions in stirred suspensions of the carboxymethyl cellulose-ficin derivative (Hornby et al., 1966; Lilly et al., 1966). The data also suggested that at very low values of Q there was a tendency for C to increase. These deviations from the behavior expected on the basis of Eq. (18) could be qualitatively explained as due to diffusion-limited transport of substrate into the enzyme particles (Hornby et al., 1968) (see also Section VI).

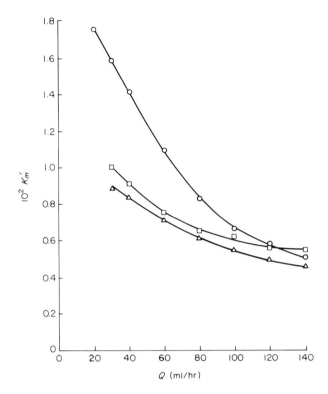

Fig. 19. Effects of flow rate, Q, on the value of the apparent Michaelis constant, K'_m, for columns of several carboxymethyl cellulose-ficin preparations acting on benzoyl-L-arginine ethyl ester. [Lilly et al. (1966).]

The efficiency of operation of biochemical reactors using immobilized enzymes in suspension in a continuous-feed stirred tank has been investigated and compared to that in a packed bed (Lilly and Sharp, 1968; Lilly et al., 1968). In a suspension of carboxymethyl cellulose-chymotrypsin in a tank agitated by a turbine impeller, both the apparent Michaelis constant and V_{max} varied with the degree of agitation, indicating that the reaction rate was partly diffusion-controlled (Lilly and Sharp, 1968; Hornby et al., 1968). Similar phenomena have been observed in studies on the rate of utilization of dissolved oxygen by microorganisms (Johnson, 1967) and of glucose by microbiological films (Atkinson et al., 1967) which were found to be limited by the rate of diffusion of substrate. Under most conditions a packed bed has been shown to be more efficient than a continuous-feed stirred tank, but diffusion limitations of the reaction could be significant in either type of reaction (Hornby et al., 1968). In a packed bed this limitation could be overcome by increasing the linear velocity of the substrate solution through the bed (Lilly and Sharp, 1968; Lilly *et al.*, 1968). One of the major problems in operating a packed bed, however, is the difficulty of getting adequate flow rates with the present enzyme support materials. Increase of the flow rate through an immobilized enzyme bed could be attained by the use of porous sheets, such as filter paper or cloth to which the enzyme has been attached. Reactive sheets of chymotrypsin, lactic dehydrogenase, creatine kinase, pyruvate kinase, and β-galactosidase have been described (Kay *et al.*, 1968; Wilson *et al.*, 1968a,b). The kinetics of enzymes attached to porous sheets appear to be similar to those of immobilized enzymes in packed beds of which they are a special case (Sharp *et al.*, 1969).

VI. ENZYME MEMBRANES

A. Structure of Enzyme-Collodion Membranes

Synthetic enzyme membranes could be prepared by adsorbing an enzyme onto preformed collodion membranes (200 μ to 500 μ thick), followed by cross-linking of the adsorbed protein (Goldman *et al.*, 1965, 1968a). Papain (Goldman *et al.*, 1965, 1968a) and alkaline phosphatase-collodion membranes (Goldman *et al.*, 1971) have been prepared by this method. Bisdiazobenzidine-2,2'-disulfonic acid and glutaraldehyde were used as cross-linking agents. The adsorption pattern could be described as saturation of successive infinitesimal layers in depth, as indicated by the sharp protein boundaries observed at different stages of the adsorption process (see Fig. 20) (Goldman *et al.*, 1968a). The thickness of the enzyme layers could be regulated by adjusting the amount of enzyme in the adsorption solution and by controlling the time of exposure of the membrane to this solution.

Three-layer enzyme membranes, consisting of two enzyme layers separated by a collodion layer, two-layer enzyme membranes consisting of an enzyme layer and a collodion layer, and a one-layer enzyme membrane (a saturated membrane) were prepared by this technique (Goldman *et al.*, 1965, 1968a). The swollen collodion membranes had an adsorption capacity of 67.5 mg/cm^3 for papain and of 70 mg/cm^3 for alkaline phosphatase. The time course of the adsorption process and the saturation values obtained with papain implied that the enzyme adsorbs as a monomolecular layer onto the surface of the pore walls of the membrane, covering about half the total available surface. Pore radii of about 300 Å were estimated for the collodion membranes (Goldman *et al.*, 1968a).

For a quantitative description of the adsorption of an enzyme on the collodion membrane it was assumed that: (a) the interstitial concentration of enzyme at the moving boundary is zero; (b) the diffusional flow is quasistationary, i.e., the

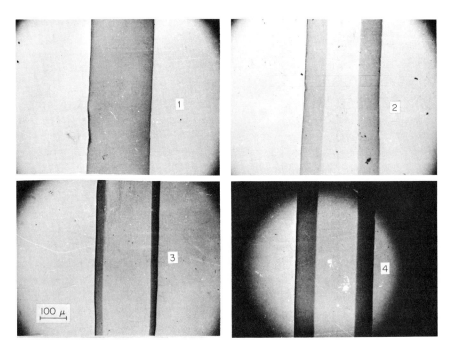

Fig. 20. Cross sections (5 μ thick) of hematoxylin-eosin stained papain collodion membranes. The papain content of the various membranes (per square centimeter of membrane) is (1), 3 mg/cm^2; (2) and (4), 1 mg/cm^2; (3), 0.3 mg/cm^2. [Goldman et al. (1968a).]

amount of enzyme entering at the solution-membrane interface is practically equal to that adsorbed per unit time; (c) the surface diffusion of the adsorbed molecules is negligible; and (d) the diffusion coefficient of the enzyme is constant and hence the concentration profile of the free enzyme in the membrane is linear.

The flow of enzyme per unit area (J_{enz}) into a swollen collodion membrane inserted into an infinite bath of enzyme at a concentration C_e is given by

$$J_{\text{enz}} = -D' \frac{dc}{dx} = D' \frac{C_e}{\Delta x} \tag{19}$$

where D' denotes the apparent diffusion coefficient of the enzyme in the membrane, and Δx is the thickness of the enzyme saturated layer at any given time t. J_{enz} can also be represented by

$$J_{\text{enz}} = \rho \frac{d(\Delta x)}{dt} = D' \frac{C_e}{\Delta x} \tag{20}$$

where ρ is the saturation capacity of the membrane with enzyme and $d(\Delta x)/dt$ gives the rate of change of thickness of the enzyme saturated layer. Integration yields for Δx

$$\Delta x = (2D'tC_e/\rho)^{1/2} \qquad (21)$$

The relation obtained resembles the corresponding expression for the average displacement of molecules by free diffusion. It should be noted, however, that in the present case the increase in the width of the enzyme layer, when compared with free diffusion, is retarded by the factor $(C_e/\rho)^{1/2}$.

The adsorption of proteins on collodion is nonspecific; both proteins and synthetic macromolecules are readily adsorbed (Gregor and Sollner, 1946; Lewis and Sollner, 1959; Hoffer and Kedem, 1968; Goldman et al., 1968a; Goldman et al., 1971; Goldman and Lenhoff, 1969). Asymmetric enzyme membranes could in principle be prepared by using solutions of different concentration or different composition on each side of the collodion membrane.

B. Dependence of Membrane Enzymic Activity on pH

The pH dependence of the rates of hydrolysis of various substrates of papain and alkaline phosphatase embedded in collodion membranes has been found to deviate considerably from that observed with the corresponding native enzyme (Goldman et al., 1965, 1968a; Goldman et al., 1971). The rate of hydrolysis of benzoyl-L-arginine ethyl ester (BAEE) by a papain membrane, when assayed in the absence of buffer, showed a monotonic increase with pH up to pH 9.6, in contradistinction to the bell-shaped pH activity profile of the native enzyme acting on the same substrate (Fig. 21). No measurements were carried out at more alkaline pH values because of the high rate or OH^- catalyzed ester hydrolysis. The enzyme membrane also showed relatively higher activities than native papain in the acid range of pH 6.0–3.0.

The enzymic hydrolysis of esters, such as BAEE, leads to the liberation of hydrogen ions in quantitative yield in the pH range at which the acid formed is fully ionized. The hydrogen ions generated

$$R_1COOR_2 \xrightarrow{\text{papain}} R_1COO^- + H^+ + R_2OH \qquad (22)$$

in the membrane lower considerably its local pH and thus lead to a completely distorted pH activity curve (see Fig. 21). The total reaction rate is given by the sum of the enzymic rates exhibited by the various consecutive infinitesimal papain layers differing in their local pH and substrate concentration. Theoretical considerations indicate that under the conditions employed for the experiments represented in Fig. 21, one may expect a pH of about 3.0–4.0 at a distance of 1 μ

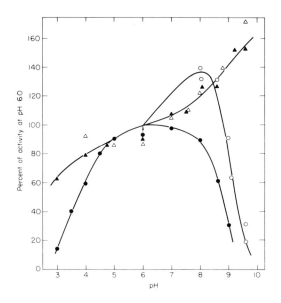

Fig. 21. pH Activity curves for papain and two papain membranes using benzoyl-L-arginine ethyl ester as substrate; ▲-▲, a three-layer papain membrane; △-△, a one-layer papain membrane; ○-○, papain membranes ground to a powder; ●-●, native papain. [Goldman *et al.* (1968a).]

from the outer surface of the papain membrane, at an external pH of 7.0. The marked increase in the activity of the membrane, as the external pH is increased from pH 7.0 to 9.6 shows that a corresponding increase in pH occurs also within the membrane. The finding that the activity of the membrane continues to increase up to an external pH of 9.6 indicates that the optimal pH of papain activity (pH 6.5–7.0) has not been attained in the membrane even at the most extreme external pH values employed.

All the cross-linked papain membranes prepared were brown. It was not possible, therefore, to determine their inner pH value by indicators. Qualitative studies on papain collodion membranes which were not cross-linked showed that their pH activity curve on BAEE resembled that of cross-linked membranes. When the neutral red indicator was added to a solution of BAEE containing an inactive papain membrane of this kind, both membrane and solution were yellow at all pH values above 7.0. On activation of the enzyme by addition of 2,3-dimercaptopropan-1-ol, the membrane became immediately red at all external pH values up to 10.0, although the indicator in solution remained yellow, implying a pH difference between the membrane and the external solution of at least 3 pH units.

The enhancement in the rate of hydrolysis of BAEE by a papain membrane on increasing the pH from 7.0 to 9.6 could be markedly diminished in the presence of external buffers of a concentration of 0.1 M. The buffers used diffuse into the membrane and increase its inner pH by neutralizing the acid generated enzymatically. Closest fit between the pH activity profile of the papain membrane and that of native papain was obtained when BAEE in 0.1 M buffer was forced through the membrane under pressure of 5 atmospheres (Fig. 22).

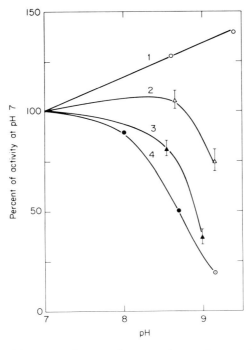

Fig. 22. pH Activity curves for a one-layer papain membrane acting on benzoyl-L-arginine ethyl ester under different conditions. (1) Papain membrane in absence of buffer; (2) papain membrane in presence of buffer; (3) papain membrane through which substrate dissolved in buffer was forced under a pressure difference of five atmospheres; (4) native papain. [Goldman et al. (1968a).]

The rates of reaction in this experiment were derived from the amount of product which appeared in the effluent. The pH of the latter was 0.2 pH units lower than that of the original substrate solution. By forcing the buffered substrate solution through the membrane, one sweeps the microenvironment of the bound enzyme with buffer and thus produces a milieu resembling that of the external solution. The finding that the pH of the effluent is practically the same as that of the original reaction mixture supports this interpretation.

The pH activity profile of particulate water-insoluble papain derivatives (Fritz et al., 1968, 1969), with BAEE as substrate, closely resembles that of native papain. It seemed relevant, therefore, to follow the change in the pH activity curve of a papain membrane on grinding the membrane to a fine powder. A papain membrane was frozen in liquid air and ground to a powder. The esteratic activity was determined as usual. Maximum activity was found at pH 8.0 (Fig. 21), the activity at pH 9.6 being only 35% of that at pH 6.0. These findings give further support to the assumption that the anomaly recorded in the pH dependence of the enzymic activity of the papain membrane is due mainly to the fact that the enzymic reaction is diffusion controlled and that no alterations in the catalytic parameters of the enzyme occurred. Benzoylglycine ethyl ester (BGEE) is hydrolyzed by papain at a considerably slower rate than BAEE. A papain membrane acting on BGEE will thus liberate, per unit time, less acid than a papain membrane acting on BAEE under the same conditions. At a stationary state this will lead to relatively higher local pH values in the membrane acting on BGEE than in the membrane acting on BAEE. One may thus expect that a high enough basic pH of the external solution, the inner pH of a papain membrane acting on BGEE will reach a pH of optimum activity (pH $\sim$ 6.0–7.0). This seems to occur at an external pH of 8.5 (see Fig. 23) where the rate of BGEE hydrolysis by the papain membrane is approximately three times greater than the rate of hydrolysis of the same substrate at pH 6.0.

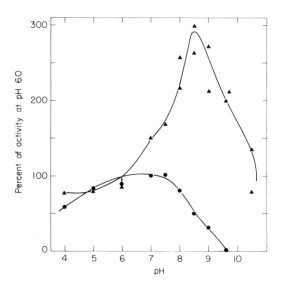

Fig. 23. pH Activity curves for a three-layer papain membrane (▲-▲) and for native papain (●-●) using benzoylglycine ethyl ester as substrate. [Goldman et al. (1968a).]

The pH activity profile of papain membranes acting on benzoyl-L-arginine amide (BAA), is bell shaped. The alkaline limb of the pH rate curve, however, is displaced toward more alkaline pH values as compared to native papain (Fig. 24). Amide hydrolysis produces carboxylate and ammonium ions in the pH range 4.5–8.0 [Eq. (23)].

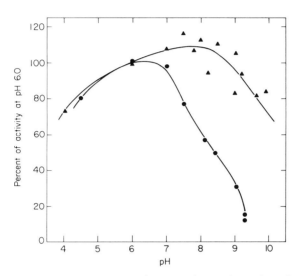

Fig. 24. pH Activity curves for a three-layer papain membrane (▲-▲) and for native papain (●-●), using benzoyl-L-arginine amide as substrate. [Goldman et al. (1968a).]

$$R_1CONH_2 \xrightarrow{papain} R_1COO^- + NH_4^+ \tag{23}$$

At this pH range one can assume that the local pH in the papain membrane equals the pH in the external solution. At higher pH values, however, the ammonium ions liberate hydrogen ions and the pH of the membrane is shifted toward more acid pH values. This would lead to a corresponding shift in the pH activity profiles toward more alkaline pH values (Fig. 24).

Similar distortions in pH activity profiles have been observed with an alkaline phosphatase-collodion membrane. The pH activity profile of the membrane bound enzyme with p-nitrophenyl phosphate as substrate displayed a continuous rise up to pH 10 (Fig. 25). The pH rate profile of native alkaline phosphatase reaches plateau values at pH's above 8.5. Complete cancellation of the pH rate anomaly of the enzyme membrane could be effected by including concentrated buffer ($\sim 0.4\ M$ borate) in the reaction mixture (Fig. 25).

It is worthwhile mentioning that comparative kinetic studies on membrane-bound and solubilized acetylcholinesterase derived from the electric tissue of *Electrophorus electricus* revealed that the particulate enzyme exhibited anomalous pH dependence of the acetylcholinesterase activity similar to that observed in

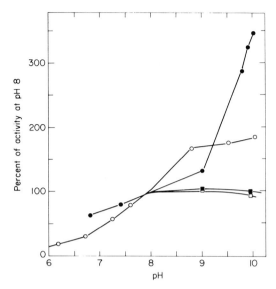

Fig. 25. pH Activity curves for an alkaline phosphatase collodion membrane and native alkaline phosphatase, using *p*-nitrophenyl phosphate as substrate in the presence of 1 M NaCl. Alkaline phosphatase membrane in the presence (■-■) and in the absence (●-●) of 0.4 M borate buffer; native alkaline phosphatase in the presence (□-□) and absence (○-○) of 0.4 M borate buffer. [Goldman et al., (1971).]

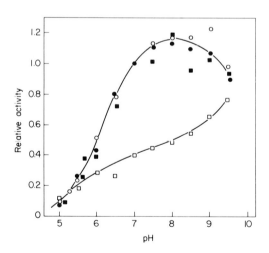

Fig. 26. pH Dependence of activity of membrane-bound acetylcholinesterase (M-AChE) and solubilized acetylcholinesterase (S-AChE). □, M-AChE, in the absence of buffer; ■, M-AChE, in the presence of buffer; ○, S-AChE, in the absence of buffer; ●, S-AChE in the presence of buffer. [Silman and Karlin (1967).]

the synthetic enzyme membrane systems described above and was probably due to the same type of local pH effects (Silman and Karlin, 1967) (Fig. 26).

Hydrogen ions are liberated or consumed in many enzymic reactions. For example, H^+ is generated in enzymic oxidation reactions in which NAD or NADP participate as cofactors in the hydrolysis of carboxylic esters, thiol esters, peptide bonds, acid anhydrides, as well as in phosphorylation reactions by ATP. Since many of these reactions occur *in vivo* in membranes or in insoluble particles, one may expect local changes in pH as a result of many localized enzymic reactions. Theoretically it could be shown that even in relatively slow reactions one might expect a difference of 2 to 3 pH units across an enzyme membrane 200–300 Å thick (Goldman *et al.*, 1965). It thus seems plausible that marked pH variations occur in biological membranes. Such variations may play an important role in determining the permeability and transport across cell membranes (Heinz, 1967) and the rate of reaction of membrane bound enzymes (Bass and McIlroy, 1968). Mitchell (1966, 1967) has suggested that pH gradients across biological membranes serve as the driving force for the formation of ATP in oxidative phosphorylation and photosynthesis (see also Energy Coupling in Electron Transport, 1967).

C. Analysis of the Kinetic Behavior of Enzyme Membranes

A membrane with enzymic activity immersed in a substrate solution will attain a stationary state within a relatively short time determined by the boundary conditions of the system (De Groot and Mazur, 1963). The stationary state of the membrane phase is characterized by the equations $(\partial S/\partial t)_{x,y,z} = 0$ and $(\partial P/\partial t)_{x,y,z} = 0$, where S and P are the local concentrations of substrate and product in the membrane. The local concentrations of substrate and product do not vary with time because of two simultaneous processes acting in opposite directions within each volume element of membrane. The disappearance of substrate, as a result of the enzymic reaction, is compensated by the net flow of substrate into the volume element as a result of diffusion. Accumulation of product is counterbalanced by the diffusion of product out of the volume element. Assuming Fick's law for the diffusion of substrate and product, the relationship between the enzymic reaction and the diffusion process in the stationary state can be summarized by Eqs. (24) and (25) (Goldman *et al.*, 1968b):

$$D'_s \frac{d^2 S}{dx^2} - f(S) = 0 \qquad (24)$$

$$D'_p \frac{d^2 P}{dx^2} + f(S) = 0 \qquad (25)$$

where $f(S)$ is the local rate of enzymic reaction and D'_s and D'_p are the apparent diffusion coefficients of substrate and product in the membrane. Henceforth, it will be assumed that the two diffusion coefficients (D'_s and D'_p) are independent of substrate and product concentration. Summation of Eqs. (24) and (25) and integration with respect to x gives:

$$D'_s(dS/dx) + D'_p(dP/dx) = a \qquad (26)$$

where a is an integration constant. $-D'_s(dS/dx) = J_s$ and $-D'_p(dP/dx) = J_p$ represent the local flows of substrate and product per square centimeter of membrane, respectively.

The relation between the local concentrations of substrate and product can be obtained by integration of Eq. (26).

$$D'_s S + D'_p P = ax + b \qquad (27)$$

The integration constants a and b are determined by the appropriate boundary conditions.

A detailed analysis of the kinetic behavior of enzyme membranes separating two infinite compartments containing substrate and product of different concentrations (see Fig. 27) has been carried out by Goldman et al. (1968b). Two simple cases will be considered below.

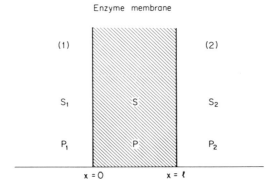

Fig. 27. Scheme describing an enzyme membrane of thickness l, separating the two compartments (1) and (2) each containing the corresponding concentrations of substrate, S, and product, P. [Goldman et al. (1968b).]

In the first case it is assumed that substrate is present only in compartment 1, and that both compartments are devoid of product. In the second case it is assumed that both compartments contain the same concentration of substrate and are devoid of product.

If we assume that the local rates of enzymic reaction in the membrane, $f(S)$, follow first order kinetics, then

1. Water-Insoluble Enzyme Derivatives

$$f(S) = kS \tag{28}$$

where k is the first order reaction rate constant. For enzyme reactions which obey Michaelis-Menten kinetics and under conditions where $K_m(\text{app}) \gg S$, the first order rate constant, k, in Eq. (28) is given by

$$k = k_{cat} E_o / K_m(\text{app}) \tag{29}$$

where E_o gives the enzyme concentration and k_{cat} is the turnover number.

Substitution of the expression for $f(S)$ [Eqs. (28) and (29)] into Eq. (24) yields a second order differential equation the solution of which is

$$S = A \exp(\alpha x) + B \exp(-\alpha x) \tag{30}$$

$$\alpha = (k/D'_s)^{1/2} \tag{31}$$

The integration constants A and B can be evaluated from Eq. (30) by introducing the boundary conditions pertinent to the case under discussion.

Case I: $S_1 \neq 0$ and $S_2 = P_1 = P_2 = 0$ (See Fig. 27)

The explicit dependence of the local substrate concentration in the membrane upon α, x and the thickness of the membrane, l, is given for this case by

$$S = \frac{S_1 \sinh \alpha (l-x)}{\sinh(\alpha l)} \tag{32}$$

The flow of substrate, J_s, at any point x is determined by the substrate concentration gradient at this point

$$\frac{dS}{dx} = -\frac{\alpha S_1 \cosh \alpha (l-x)}{\sinh(\alpha l)} \tag{33}$$

In a system exposed to the asymmetric boundary conditions under discussion substrate flows unidirectionally through the membrane from compartment 1 to compartment 2. The flow of product on the other hand, varies in direction, product being liberated into both compartments.

The local concentration of product, P, and the product concentration gradient, dP/dx, can be derived from Eqs. (26), (27), (32), and (33):

$$P = -\frac{D'_s S_1}{D'_p} \left[\frac{\sinh \alpha(l-x)}{\sinh(\alpha l)} + \frac{x}{l} - 1 \right] \tag{34}$$

$$\frac{dP}{dx} = \frac{D'_s S_1}{D'_p} \left[\frac{\alpha \cosh \alpha(l-x)}{\sinh(\alpha l)} - \frac{1}{l} \right] \tag{35}$$

The substrate and product concentration profiles in an enzymically active membrane, calculated from Eqs. (32) and (34), are shown in Fig. 28. The curves

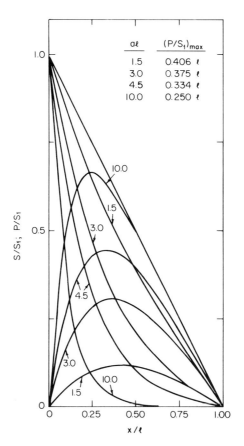

Fig. 28. Calculated concentration profiles for substrate and product in an enzyme membrane exposed to the asymmetric boundary conditions specified for Case I (see text). The local concentration of substrate, S, was calculated by making use of Eq. (32); the local concentration of product, P, was calculated with the aid of Eq. (34) assuming $D'_s = D'_p$. The arbitrary αl values chosen for the different curves presented are specified on the curves. The local concentrations of substrate and product within the membrane are expressed as fractions of the concentration of substrate, S_1, at $x = 0$. [Goldman et al. (1968b).]

for the different αl values were calculated, assuming the equality of diffusion coefficient of substrate and product, i.e., $D'_s = D'_p$. The substrate concentration profiles reveal that at any given value of x, S decreases with increasing αl. Thus on increasing αl the thickness of the enzyme layer participating in the catalytic reaction decreases, whereas the thickness of the layer devoid of substrate would increase. The product concentration profile passes through a maximum value whose coordinates are determined by αl. An increase in αl leads to a concomitant increase in the value of P_{max} and a decrease in the value of x at which P_{max} appears.

The ratio of the flows of product at $x=0$, J_p^0 and at $x=l$, J_p^l, is given in Eq. (36).

$$J_p^0/J_p^l = \frac{\alpha l \cosh(\alpha l) - \sinh(\alpha l)}{\alpha l - \sinh(\alpha l)} \tag{36}$$

The ratio J_p^0/J_p^l is thus determined only by the activity of the immobilized enzyme and by the physical parameters of the membrane (D_s', E_0, and l), and is independent of the concentration of substrate in compartment 1.

Equation (36) shows that for enzyme membranes of high activity for which $\alpha l \gg 4$ and therefore $\sinh(\alpha l) \approx \cosh(\alpha l) \gg \alpha l$, $J_p^0/J_p^l = \alpha l - 1$; i.e., the flow of product into compartment 1 markedly exceeds that into compartment 2. It is of interest that even for enzyme membranes of low activity ($\alpha l \to 0$), $J_p^0/J_p^l = 2$.

The overall rate of substrate consumption or product formation per square centimeter of an enzyme membrane, V, exposed to the boundary conditions of Case I is given by the difference between the flows of substrate at $x=0$ and $x=l$, or by the sum of product flows out of the membrane:

$$V = J_p^l - J_p^0 = J_s^0 - J_s^l = \alpha S_1 D_s'[\cosh(\alpha l) - 1]/\sinh(\alpha l) \tag{37}$$

Case II: $S_1 = S_2 = S_0$ and $P_1 = P_2 = 0$ (See Fig. 27)

The dependence of local concentrations of substrate on x under symmetric boundary conditions is given by

$$S = \frac{S_0[\sinh(\alpha x) + \sinh \alpha(l-x)]}{\sinh(\alpha l)} \tag{38}$$

It follows from Eq. (27) that at any point in the membrane for symmetric boundary conditions (with $D_s' = D_p'$)

$$S + P = S_0 \tag{39}$$

Under symmetric boundary conditions, the concentration profiles of substrate and product are thus complementary (Fig. 29).

Figure 29 shows that the substrate concentration approaches zero at the midpoint of the membrane, whereas product concentration approaches S_0, as αl increases from 1.5 to 10. The concentration of substrate decreases gradually toward the inner parts of the membrane for any given value of αl, and reaches minimum at $x/l = 0.5$. The calculations further show that for membranes with

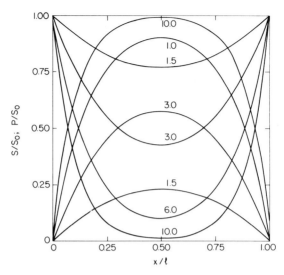

Fig. 29. Calculated concentration profiles for substrate and product in an enzyme membrane exposed to the symmetric boundary conditions specified for Case II (see text). The local concentration of substrate, S, was calculated by means of Eq. (38). The local concentration of product, P, was calculated by means of Eq. (39). $D'_s = D'_p$ and $P_0 = 0$ were assumed. The arbitrary αl values chosen for the different curves presented are specified on the curves. The local concentrations of substrate and product within the membrane are expressed as fractions of the concentration of substrate at the external solution, S_0. [Goldman et al. (1968b).]

high enzymic activity ($\alpha l \gg 4$), the volume fraction of the membrane which participates in the enzymic reaction decreases on increasing the specific enzymic activity (α) of the membrane.

The overall rate of an enzyme membrane catalyzed reaction, V, is given in this case by:

$$V = 2J_s^0 = \frac{2\alpha S_0 D'_s[\cosh(\alpha l) - 1]}{\sinh(\alpha l)} \tag{40}$$

At high αl values ($\alpha l \gg 4$), Eq. (40), reduces to

$$V = 2D'_s \alpha S_0 \tag{41}$$

The overall reaction rate of a membrane with high enzymic activity is therefore independent of the thickness of the enzyme layer, and is directly proportional to the external concentration of substrate.

Comparison of the overall rate of an enzyme membrane, V, to that of an equal amount of soluble enzyme, V_0, can serve as a measure for the effectiveness of the immobilized enzyme.

The effectiveness factor f is defined by

$$f = V/V_o = \frac{2[\cosh(\alpha l) - 1]}{\alpha l \sinh(\alpha l)} \tag{42}$$

V_0, the overall rate of the soluble enzyme, is given by

$$V_0 = klS_0 \tag{43}$$

The plot of f vs. αl calculated according to Eq. (42) (Fig. 30) shows that f decreases markedly on increasing αl. At very low αl values ($f \to 1$) the activity of the membrane-bound enzyme equals that of an equivalent amount of native enzyme.

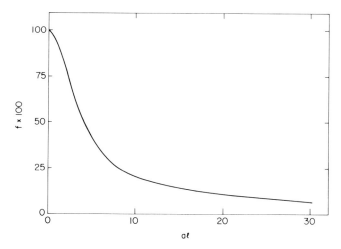

Fig. 30. Variation of the parameter f with αl. $f = V/V_o$, where V is the rate of reaction of an enzyme membrane exposed to the boundary conditions specified for Case II, and V_o is the rate of reaction of the soluble native enzyme in an amount equal to that present within the membrane when exposed to similar conditions. The curve was calculated by means of Eq. (42). [Goldman et al. (1968b).]

The validity of the conclusions drawn from Eq. (42) could be tested by measuring the rates of hydrolysis of a poor substrate such as acetyl-L-glutamic acid diamide (AGDA) (Blumenthal et al., 1967) and a good substrate such as benzoyl-L-arginine amide (BAA) (Whitaker and Bender, 1965), which differ markedly in α, by three papain collodion membranes of an equivalent enzyme membrane thickness of 470 μ, 156 μ, and 49 μ. The three membranes tested on AGDA showed a linear increase in activity with increase in enzyme membrane thickness. This is in accord with the fact that for these enzyme membrane substrate systems $\alpha l < 1$. For the 470 μ and 156 μ membranes acting on BAA, $\alpha l > 4$. Both membranes thus showed similar activities toward BAA, despite the difference in their thickness. The 49 μ membrane, however, for which $\alpha l = 1.77$, hydrolyzed BAA at a lower rate than the 470 μ and 156 μ membranes.

A similar approach has been used to characterize heterogeneous catalysis by porous catalysts (both organic and inorganic) in processes which proceed by first order kinetics. The "degree of catalyst utilization," a variable analogous to f, was found to depend upon the "Thiele modulus," a variable analogous to αl (Thiele, 1939; Wheeler, 1951; Helfferich, 1962b).

The recent advances in crystallography have brought forth the need to evaluate the catalytic activity of enzyme molecules assembled in a crystal lattice. The full activity of an enzyme crystal cannot be realized in most cases due to diffusional rate-limiting effects.

The effectiveness factor, f, of an enzyme crystal can be calculated from the kinetic parameters of the native enzyme in solution [k_{cat}, $K_m(\text{app})$], the thickness of the crystal [a parameter analogous to l in Eq. (42)] and the diffusion coefficient of substrate in the crystal (D'_s) [c.f. Eq. (42)]. The actual amount of active enzyme in the crystal can be obtained by dividing the experimentally determined activity by the effectiveness factor.

Sluyterman and De Graaf (1969) have defined a critical size of a crystal, d_c, as the size at which the effectiveness factor is 0.92 (i.e. the enzyme in the crystal exhibits 92% of its activity in solution). The parameter d_c is given by

$$d_c = [(K_m(\text{app}) D'_s / k_{cat} E_0]^{1/2} \qquad (44)$$

It can be easily seen that $d_c = 1/\alpha$ using our notation. Critical size conditions can be attained [c.f. Eq. (44)] by adjusting either the kinetic parameters [$K_m(\text{app})/k_{cat}$] by the appropriate choice of substrate, or by changing the concentration of active enzyme in the crystal. The validity of Eq. (44) could be tested by means of two types of papain crystals for which $\alpha l = 1$, using benzoyl-L-arginine ethyl ester (a good substrate) and acetylglycine ethyl ester (a poor substrate) (Sluyterman and De Graaf, 1969). The two crystalline modifications of papain exhibited full activity with either substrate. This is the first case reported in the literature in which complete activity of an enzyme crystal could be demonstrated in a straightforward manner.

Studies on the apparent activity of several enzyme crystals have been reported. Reasonably high activity was found in crystalline ribonuclease-S (Doscher and Richards, 1963) (5–25%), and in chymotrypsin (Kallos, 1964) (20%). On the other hand, very low reactivity was recorded for crystalline carboxypeptidase (Quiocho and Richards, 1966a) (0.3%) and for liver alcohol dehydrogenase (Theorell et al., 1966) (0.1%). The above considerations (the effectiveness factor) when applied to chymotrypsin and ribonuclease-S have shown that both enzymes, like papain, very possibly exhibit full activity in the crystalline state (Sluyterman and De Graaf, 1969).

So far we have dealt in some detail with the behavior of enzyme membranes obeying locally first order kinetics. In the following the more general case in which the enzyme membrane obeys locally the Michaelis-Menten kinetics will be discussed. The dependence of the apparent Michaelis constant on local substrate and product concentrations will be derived for the boundary conditions specified for Case II.

Assuming that $-f(S)$ obeys the Michaelis-Menten kinetics, the stationary state requires

$$D'_s \frac{d^2 S}{dx^2} = \frac{k_{cat} E_0 S}{K_m(\text{app}) + S} \qquad (45)$$

Integration of Eq. (45) between any given point (x) in the membrane and the midpoint ($l/2$) yields for the concentration gradient

$$\frac{dS}{dx} = \left\{ C \left[S - S' + K_m(\text{app}) \ln \frac{K_m(\text{app}) + S'}{K_m(\text{app}) + S} \right] \right\}^{1/2} \qquad (46)$$

where S' is the substrate concentration at $x = l/2$, and $C = 2k_{cat} E_0 / D'_s$.

The explicit dependence of substrate concentration S upon x cannot be derived readily from Eq. (46). The value of S' for any given value of S_0 can be obtained, however, by numerical integration of Eq. (47).

$$C^{-1/2} \int_{S_0}^{S'} \left[S - S' + K_m(\text{app}) \ln \frac{K_m(\text{app}) + S'}{K_m(\text{app}) + S} \right]^{-1/2} ds = l/2 \qquad (47)$$

Introducing the values of S', as calculated by means of Eq. (47) into Eq. (48), allows the calculation of the overall rate, V, of the enzymic reactions.

$$V = -2D'_s \left(\frac{ds}{dx} \right)_0 = - \left\{ 8D'_s k_{cat} E_0 \left[S_0 - S' + K_m(\text{app}) \ln \frac{K_m(\text{app}) + S'}{K_m(\text{app}) + S_0} \right] \right\}^{1/2} \qquad (48)$$

The dependence of V of an enzyme membrane on S_0 and the kinetic parameters of the enzyme is illustrated below for two hypothetical enzyme membrane systems with k_{cat} and K_m values corresponding to those of papain (Goldman et al., 1968b) and of alkaline phosphatase (Goldman et al., 1971). The two membrane systems were chosen so that they had similar membrane parameters.

The variation of V with S_0 for two hypothetical papain membranes of thickness 100 μ and 200 μ, acting on BAA, is shown in Fig. 31. The normalized curves calculated for the two membranes are displaced with respect to the curve calculated for the native enzyme. Whereas the native enzyme attains its half-maximal rate at $S_0 = 0.032\ M$ [the $K_m(\text{app})$ of native papain], the papain membranes attain $V/V_{max} = 0.5$ at higher values of $S_0 (S_0 = 0.045\ M$ for the 100 μ membrane; $S_0 = 0.08\ M$ for the 200 μ membrane). An apparent Michaelis constant [$K'_m(\text{app})$] of an enzyme membrane (the value of S_0 at which the enzyme membrane attains half-maximal rate) is thus higher than $K_m(\text{app})$ of the native enzyme.

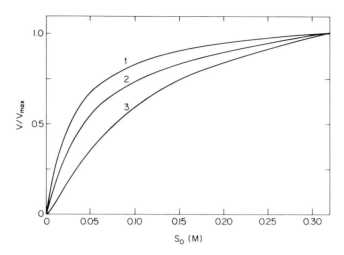

Fig. 31. Normalized rates of hydrolysis of benzoyl-L-arginine amide by native papain (curve 1) and by one-layer papain membranes, 100 μ and 200 μ thick (curves 2 and 3, respectively), as a function of the external substrate concentration, S_0. The substrate concentration at the midpoint of the membrane, S', and the reaction rate, V, were calculated by means of Eqs. (47) and (48), respectively; V, for native papain was calculated from the Michaelis-Menten relation [Eq. (4)]. The value $k_{cat} = 8.7$ sec^{-1} (Whitaker and Bender, 1965), $K_m(app) = 0.032$ M (Whitaker and Bender, 1965), $E_0 = 1.4$ mM (Goldman et al., 1968b) and $D'_s = 3 \times 10^{-6}$ cm^2 sec^{-1} (Goldman et al., 1968a) were used. [Goldman et al. (1968b).]

The calculated V vs. S_0 curves for the hypothetical alkaline phosphatase membranes of thickness 1.5, 3.0, and 9.0 μ are given in Fig. 32. The $K'_m(app)$ values obtained from curves 1,2, and 3 are 0.55×10^{-4} M, 1.2×10^{-4} M, and 6.8×10^{-4} M, respectively.

In the mathematical analysis given above product inhibition was not considered. An estimate of the effect of local product inhibition on the overall reaction rate, V, and on the value of $K'_m(app)$ could be obtained by an analysis of the kinetic behavior of an alkaline phosphatase membrane catalyzed hydrolysis of p-nitrophenyl phosphate (PNPP) under the symmetric boundary conditions of Case II (Goldman et al., 1971).

On the assumption that for this system $D'_s = 2D'_s$, i.e., the diffusion coefficient of phosphate in the membrane is twice that of PNPP, the relation between S and P at any given point x is given by Eq. (49) derived from Eq. (27).

$$S + 2P = S_0 \tag{49}$$

Since phosphate is a competitive inhibitor of alkaline phosphatase (Garen and Levinthal, 1960), the local reaction rate is described by the expression:

$$V = f(S) = \frac{k_{cat} E_0 S}{K_m(\text{app}) \left(1 + \dfrac{P}{K_I}\right) + S} \quad (50)$$

where P stands for the concentration of phosphate and K_I for the inhibition constant. Equations (24), (49), and (50) yield:

$$D'_s \frac{d^2 S}{dx^2} = \frac{2 K_I k_{cat} E_0 S}{[2 K_I K_m(\text{app}) + K_m(\text{app}) S_0] + [2 K_I - K_m(\text{app})] S} \quad (51)$$

The relation between S' and $l/2$ for any given value of S_0 is given for the present case by

$$\left(\frac{2C}{B^2}\right)^{-1/2} \int_{S_0}^{S'} \left[B(S - S') + A \ln \frac{A + BS'}{A + BS} \right]^{-1/2} ds = l/2 \quad (52)$$

where

$$A = 2 K_I K_m(\text{app}) + K_m(\text{app}) S_0$$
$$B = 2 K_I - K_m(\text{app})$$
$$C = 2 K_I k_{cat} E_0 / D'_s$$

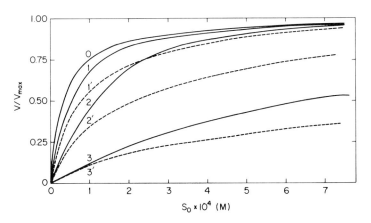

Fig. 32. Normalized rates of hydrolysis of *p*-nitrophenyl phosphate by native alkaline phosphatase and by alkaline phosphatase membranes as a function of the external substrate concentration, S_0. Curve 0, native alkaline phosphatase [calculated from the Michaelis–Menten relation (Eq. 4)]. Curves 1–3, alkaline phosphatase membranes, product inhibition neglected [calculated according to Eqs. (47) and (48)]. Curves 1'–3', alkaline phosphatase membranes, product inhibition included [calculated according to Eqs. (52) and (53)]. The values $k_{cat} = 63$ sec^{-1} (Trentham and Gutfreund, 1968), $K_m(\text{app}) = 3.4 \times 10^{-5}$ M, $E_0 = 0.82$ mM (Goldman *et al.*, 1971), $D'_s = 2.3 \times 10^{-6}$ cm^2 sec^{-1} (Goldman and Katchalski, 1971), $D'_p = 2 D'_s$ (Goldman *et al.*, 1971), and $K_I = 0.32 K_m(\text{app})$ (Garen and Levinthal, 1960; Goldman *et al.*, 1971) were used. [Goldman *et al.*, (1971).]

The overall rate of the reaction, V, is

$$V = -2D'_s \left(\frac{dS}{dx}\right)_0 = -\left\{\frac{8D_s'^2 C}{B^2}\left[B(S_0 - S') + A \ln \frac{A + BS'}{A + BS_0}\right]\right\}^{1/2} \tag{53}$$

Equations (52) and (53) can be derived from Eq. (51) by a procedure analogous to the one employed for the derivation of Eqs. (47) and (48).

The calculated V vs. S_0 curves of the alkaline phosphatase membranes for which product inhibition has been taken into account are shown in Fig. 32. The values of $K'_m(\text{app})$ obtained from curves 1′, 2′, and 3′ of Fig. 32 are 0.8×10^{-4} M, 2.2×10^{-4} M and 18×10^{-4} M, respectively. The data presented in Fig. 32 shows that product inhibition in an enzyme membrane effects an increase in the values of $K'_m(\text{app})$ by a factor of about two as compared with the $K'_m(\text{app})$ values obtained in a similar enzyme membrane devoid of product inhibition.

Comparison of the computed properties of the two enzyme membrane systems dealt with, the papain and the alkaline phosphatase hypothetical membranes, shows that when diffusion is rate limiting an increase in the values of the apparent Michaelis constants is to be expected. For each membrane system an increase in the membrane thickness leads to an increase in the value of $K'_m(\text{app})$. This effect is considerably larger in the alkaline phosphatase membrane than in the papain membrane although the former membrane is about a hundred times thinner.

Since the values of D'_s and E_0 for both systems are rather similar, these differences in kinetic behavior can be attributed to differences in the values of the kinetic parameters [k_{cat} and $K_m(\text{app})$] of the enzymes themselves, alkaline phosphatase being a much more active enzyme than papain.

Finally, an additional effect, which may be of importance when the theoretical model described above is applied to experiment, should be considered.

The main difference between the mathematical model of an enzyme membrane system and real enzyme membranes resides in the fact that in the model the existence of unstirred layers (referred to as "Nernst diffusion layers") was ignored. The existence of such layers is well known from hydrodynamic studies and was invoked as early as 1904 by Nernst (1904) to explain the behavior of heterogeneous catalysis. Estimates of the thickness of such diffusion layers, which are independent of both the solute and the membrane but depend on the conditions of shaking or stirring, range from 10–100 μ (Helfferich, 1962a; Ginzburg and Katchalsky, 1963). An estimate of the Nernst diffusion layer adhering to an enzyme membrane could be obtained using the following argument (Goldman et al., 1971). In the stationary state the enzymic reaction establishes concentration gradients of substrate across the Nernst diffusion layer. The flow of substrate, J_s, through the unstirred layer is given by

$$J_s = -D_s(S'_o - S_0)/\delta \tag{54}$$

where S_0 and S'_o are the substrate concentration in the bulk of the solution and in contact with the enzyme membrane, i.e., at $x=0$ and $x=l$, respectively. D_s is the diffusion coefficient of substrate in the solution, and δ is the thickness of the unstirred layer. In the stationary state the net flow of substrate through the unstirred layers equals the rate of its consumption in the membrane by the enzymic reaction.

At symmetric boundary conditions

$$V = 2J_s = 2D_s(S_0 - S'_o)/\delta \tag{55}$$

The values of S'_o at a given value of S_0 strongly depend on the catalytic parameters of the enzyme substrate system. High activity of the enzyme layer leads to low values of S'_o, i.e., high values of S_0 will be needed in order to reach $V_{\max}$. The values of the Michaelis constants derived experimentally for an enzyme membrane [K'_m(app)] will therefore differ considerably from those computed for the model system where effects of the unstirred layers were ignored.

An estimate of the thickness of the unstirred layer, δ, can be obtained by inserting into Eq. (55) the experimental values of $V_{\max}/2$ and S_0 and of the corresponding calculated values of K'_m(app) (including product inhibition corrections) for S'_o (see Fig. 32). Values of δ ranging from 42 μ to 66 μ were obtained by this procedure for several alkaline phosphatase collodion membranes (Goldman et al., 1971).

The above theoretical treatment has been recently extended by Sundaram et al. (1970) who analyzed in detail the effect of partitioning of substrate between the support and the free solution and of diffusion of substrate in the support on the overall rate of enzymic reactions.

VII. APPLICATIONS

Immobilized enzymes in particulate form, in suspension or in packed columns, as well as enzyme membranes, can be used in continuous catalytic processes, in controlling substrate conversion, in the detection and assay of the appropriate compounds, and as specific adsorbents. Of the variety of immobilized enzymes described in the literature (see Table I), the hydrolyzing enzymes, particularly the proteases, have received by far the widest attention. The applications to be described below, though mainly of academic interest, suffice to show that a new technology is being developed.

Water-insoluble papain was utilized in the partial degradation of rabbit γ-globulin by Cebra et al. (1961, 1962; Cebra, 1964; Jaquet and Cebra, 1965). The water-insoluble enzyme was activated with cysteine prior to use, and the excess cysteine removed by washing without impairing its proteolytic activity (Cebra et al., 1961). Brief incubation of rabbit γ-globulin with cysteine-free activated water-insoluble papain and removal of the enzyme after only three to five peptide bonds had been split caused no appreciable change in the sedimentation coefficient of the native protein (6.25 S). However, when the γ-globulin, thus treated, was incubated with cysteine after removal of the insoluble enzyme, it was completely transformed into the 3.3 S fragments (F_{ab} and F_c) similar to those described by Porter (1959). These results showed that the cleavage of rabbit γ-globulin by cysteine-activated native papain, as demonstrated by Porter (1959), consists of two discrete stages: proteolysis followed by reduction. If instead of normal rabbit γ-globulin, rabbit antiovalbumin was briefly digested with activated water-insoluble papain in the absence of cysteine, its ability to precipitate its specific antigen was found to be unimpaired (Cebra et al., 1961). However, when immune precipitates so obtained were treated with cysteine or thioglycolic acid, complete dissolution occurred, whereas precipitates formed with intact γ-globulin were unaffected. The solubilized precipitate gave two sedimentation peaks in the ultracentrifuge; the rapidly sedimenting peak was shown to consist of one antigen molecule combined with several monovalent antibody F_{ab} fragments, whereas the slowly sedimenting peak was shown to consist of the F_c fragment, which is devoid of antibody activity. The antigen-antibody fragment complex could be chromatographically separated from the immunologically inert fragment (Cebra et al., 1962). Complexes analogous to the above were also obtained from precipitates derived from the reaction of lysozyme and bovine serum albumin with their corresponding rabbit antibodies which were pretreated with water-insoluble papain (Cebra et al., 1962). More recently, Cebra (1964; Jaquet and Cebra, 1965) has shown that if the brief digestion of antibody rabbit γ-globulin with water-insoluble papain is followed by treatment with sodium dodecylsulfate in the absence of reducing agents, a product denoted as the F_{ab} fragment dimer is obtained, resembling in its physicochemical and immunochemical properties the "pepsin product" which Nisonoff and his co-workers (1960a,b) obtained by exhaustive pepsin proteolysis of rabbit antibody.

Two water-insoluble trypsin derivatives, insoluble polytyrosyl trypsin and the polyanionic, ethylene-maleic acid copolymer-trypsin conjugate (EMA-trypsin), were used by Alexander et al. (1965) in the investigation of the early stages of tryptic digestion of fibrinogen. Both derivatives were found to impair the clottability of fibrinogen by thrombin similarly to native trypsin. Complete incoagulability ensued when an average of eight peptide bonds per fibrinogen molecule had been cleaved by either of the two insoluble trypsin derivatives. Scission of one to two bonds, however, substantially increased the time of clotting by thrombin. Limited digestion of purified fibrinogen using a water-insoluble polyanionic derivative of thrombin (EMA-thrombin) led to the formation of a cold-insoluble, highly clottable form of fibrinogen-cryofibrinogen

(Cohen et al., 1966). Cohen et al. (1966) have shown that cryofibrinogen tends to form periodic aggregates, resembling those of fibrin, whereas precipitates from cold-soluble fibrinogen have a different morphology.

The use of water-insoluble enzymes in the activation of zymogens without contaminating the system with the activating enzyme may be illustrated by the conversion of chymotrypsinogen to chymotrypsin using a carboxymethyl cellulose conjugate of trypsin (Mitz and Summaria, 1961), insoluble maleic acid-ethylene trypsin (Levin et al., 1964), or insoluble polytyrosyl trypsin (Levin et al., 1964).

Several cases of modified hydrolysis patterns presumably due to restrictions imposed on the specificity of immobilized proteases by the carrier have been reported in the literature (Levin et al., 1964; Alexander et al., 1965; Ong et al., 1966; Cohen et al., 1966; Lowey et al., 1966, 1967, 1968; Slayter and Lowey, 1967; Lowey, 1968; Hornby et al., 1968; Lilly et al., 1968; Westman, 1969). Both water-insoluble polytyrosyl trypsin and EMA-trypsin attacked prothrombin. The polytyrosyl trypsin derivative, as native trypsin, converts prothrombin to thrombin which is not digested any further. EMA-trypsin, however, rapidly degrades the newly formed thrombin (Alexander et al., 1966). It has also been found that the clotting factor VII, which is activated by insoluble polytyrosyl trypsin, is not affected by EMA-trypsin (Silman and Katchalski, 1966). Ong et al. (1966) have shown that whereas trypsin hydrolyzes 15 lysyl peptide bonds of pepsinogen, the maximal number of bonds cleaved by EMA-trypsin does not exceed 10. The same difference was observed when reduced carboxymethylated pepsinogen was used as substrate. These findings were confirmed by peptide mapping.

Lowey et al. (1966, 1967; Lowey, 1968; Slayter and Lowey, 1967) investigated the digestion of myosin and of the meromyosins by trypsin and by the polyanionic EMA-trypsin. The first order rate constants estimated for the EMA-trypsin digestion of myosin were about fiftyfold lower as compared to those of native trypsin. Moreover, about half as many peptide bonds were found to be susceptible to hydrolysis by the polyanionic trypsin derivative. The protein fragments obtained on limited digestion of heavy meromyosin (HMM) and myosin with EMA-trypsin differed from those obtained with the native enzyme (Lowey et al., 1966, 1967). The digestion of myosin with native trypsin leads to the formation of light and heavy meromyosins (LMM and HMM) as the only identifiable degradation products at the early stages of the reaction. Tryptic digestion of purified HMM yields the globular section of the heavy meromyosin molecule (HMM-subfragment-1), possessing unimpaired ATPase activity, whereas the rodlike helical section is degraded into peptide fragments. In contradistinction, on digesting HMM with EMA-trypsin, the globular part is primarily attacked, leaving the helical rod (HMM-subfragment-2) essentially intact.

Controlled digestion of myosin with a water-insoluble papain preparation showed that the globular, enzymic part of myosin, HMM-subfragment-1, can be split from the structural helical rod, leaving the latter intact along its length (Kominz et al., 1965; Slayter and Lowey, 1967; Lowey, 1968; Lowey et al., 1968). Further digestion cleaves the rod along two-thirds of its length to form light meromyosin and the rod portion of heavy meromyosin, HMM-subfragment-2 (Lowey et al., 1968). Hydrodynamic and electron microscopic characterization of the purified fragments led to the currently accepted model of the myosin molecule: a two-coiled-coils helical rod terminating in two globules (Slayter and Lowey, 1967; Lowey, 1968; Lowey et al., 1968).

The observations described above indicate that the carrier, at least in the case of polyelectrolyte enzyme derivatives, imposes characteristic restrictions on the specificity of a bound enzyme acting on a high molecular weight substrate. The restricted specificity is probably due to steric effects and to charge interactions between the carrier and different regions or different amino acid sequences on the polymeric substrate. A series of derivatives of a single enzyme could thus be utilized for the fragmentation of a protein into peptides of overlapping sequences.

A water-insoluble trypsin (ethylene-maleic acid copolymer) conjugate (EMA-trypsin, see Fig. 4), packed in a column, was used by Fritz et al. (1966, 1967, 1968, 1969) for the selective adsorption, at neutral pH, of trypsin inhibitors from crude extracts of animal tissue and body fluids. The inhibitors were eluted from the column with pH 2 buffers of moderate ionic strength. The inhibitor binding capacity of EMA-trypsin was about 20% of the value expected on the basis of the amount of immobilized enzyme. At 4° and in the presence of added bactericides, no signficant loss of inhibitor binding capacity was observed after more than a hundred operations in the course of 10 months.

EMA-trypsin and EMA-chymotrypsin columns were used for the separation and purification of the various protease inhibitors present in animal tissue. Bovine pancreas, for example, contains two types of protease inhibitors: (1) a nonspecific inhibitor, inhibiting trypsin, chymotrypsin, plasmin, and kallikrein, usually designated as the "Kunitz inhibitor," or the "trypsin-kallikrein inhibitor," and (2) a specific inhibitor inhibiting trypsin exclusively. Both inhibitors are basic polypeptides and have a molecular weight of about 6000. Fritz et al. (1966, 1967, 1968, 1969) could isolate and purify these inhibitors using the following procedure: the bovine pancreas extract was passed through an EMA-chymotrypsin column. Only the nonspecific Kunitz inhibitor was retained on the column, whereas the specific trypsin inhibitor passed through. About 99% of the total protease inhibitor content of the eluate was found to consist of the specific trypsin inhibitor. Purification of this inhibitor was effected by readsorption on an EMA-trypsin column, elution, and recycling of the eluent through an EMA-chymotrypsin column to remove traces of the contaminating Kunitz inhibitor still present. The nonspecific inhibitor could be purified similarly, using an EMA-kallikrein column.

Reversal of the above procedure, i.e., binding of the trypsin-kallikrein inhibitor to EMA, allowed the purification of trypsin and chymotrypsin.

The polyanionic EMA-trypsin conjugates were found to be unsuitable for the binding of the high molecular weight, acidic protease inhibitors such as soybean trypsin inhibitor or egg white trypsin inhibitor possessing an isoelectric point of 5 or below. These inhibitors could be isolated and purified, however, by their specific adsorption onto EMA-trypsin conjugate in which the negative charge of the carrier carboxyl groups was partially neutralized by the procedure outlined in Fig. 5. The methods described above have been used so far for the isolation and purification of the following protease inhibitors: trypsin inhibitors from swine, bovine and sheep pancrease; trypsin inhibitors from soybean and ovomucoid; trypsin-plasmin inhibitor from the seminal vesicles of the guinea pig; trypsin-kallikrein inhibitor from bovine lung; and trypsin inhibitors from swine sera.

Using a similar approach Givol et al. (1970) have isolated an "active site peptide" from bovine pancreatic ribonuclease digests. Ribonuclease was labeled chemically, with the active site directed reagent derived from 5'-(4-aminophenylphosphoryl)uridine 2'(3')-phosphate (PUDP) by diazotization. The labeled enzyme was reduced, carboxymethylated, and digested with trypsin. The tryptic digest was passed through a column containing a ribonuclease-Sepharose conjugate. The uridine diphosphate-labeled peptide was specifically adsorbed on the enzyme column and could be eluted with $0.8\ M$ NH_4OH. This method which is of general applicability makes use of the affinity of the native enzyme to a ligand which remains linked to the active site peptide after partial digestion of the corresponding labeled protein.

In this connection it is pertinent to note that the inherent binding capacity of enzymes toward their specific substrates or inhibitors form the basis for a new method of enzyme purification recently referred to as "affinity chromatography" (Cuatrecasas et al., 1968; Baker, 1967). (Also see the article by Cuatrecasas in this book.) In this method a crude enzyme preparation is passed through a column containing a resin to which a specific competitive inhibitor of the enzyme has been covalently attached. Proteins of no substantial affinity for the bound inhibitor pass directly through the column, whereas the enzyme which recognizes the inhibitor is retarded in proportion to its affinity constant. Elution of the bound enzyme can be effected by changing such parameters as salt concentration, pH, or addition of a low molecular weight competitive inhibitor. Affinity chromatography procedures for the purification of staphylococcal nuclease (Cuatrecasas et al., 1968), α-chymotrypsin (Wilchek, 1969), carboxypeptidase A (Wilchek, 1969), and papain (Blumberg et al., 1969, 1970) have been recently reported.

Water-insoluble antigen derivatives have been used extensively, in a similar fashion, for the isolation and purification of the corresponding homologous antibodies from immune sera (Silman and Katchalski, 1966; Sehon, 1967).

The application of immobilized enzymes in biochemical analysis is still in its initial stages. The examples cited below were chosen to illustrate the inherent potential of the new catalytic reagents.

Columns of water-insoluble urease have been used by Riesel and Katchalski (1964) to assay urea in biological fluids such as urine and serum. The ammonia liberated was determined colorimetrically. Horseradish peroxidase coupled to CM-cellulose has been used for the estimation of trace amounts of hydrogen peroxide (Weetall and Weliky, 1966; Weliky et al., 1969).

The preparation of immobilized cholinesterase for use in analytical chemistry was described by Bauman et al. (1965, 1967). The enzyme, immobilized by the use of a starch matrix and placed on a polyurethane foam pad, was found to be stable and active for 12 hours under continuous use. The activity of the enzyme was monitored electrochemically, using two platinum electrodes and an applied current of 2 μA. This immobilized enzyme was used to determine the substrates acetyl and butyryl thiocholine iodide, both in individual samples and continuously. A fluorometric system for the assay of anticholinesterase compounds using this immobilized chloinesterase with 2-naphthyl acetate as substrate was described by Guilbault and Kramer (1965). As long as the enzyme is active a fluorescence is produced because of the hydrolysis of 2-naphthyl acetate to 2-naphthol. Upon inhibition, the fluorescence drops to a value approaching zero.

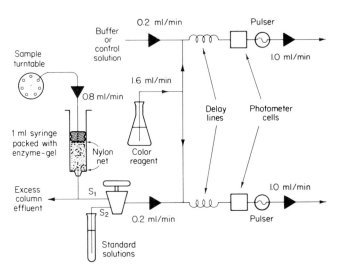

Fig. 33. Diagram of an apparatus for continuous automated analysis utilizing immobilized enzyme columns. [Hicks and Updike (1966).]

Hicks and Updike (1966) have employed immobilized enzyme columns in automated analytical procedures. The apparatus used is shown in Fig. 33. The enzymes glucose oxidase (GO) and lactic dehydrogenase (LDH) were immobilized by entrapping in polyacrylamide gel. The detection of immobilized dehydrogenase activity was based on the following reactions:

$$S + NAD \xrightleftharpoons{DH_1} P + NADH_2 \tag{56}$$

$$NADH_2 + Dye_{ox} \xrightarrow{PMS} NAD + Dye_{red} \tag{57}$$
$$\text{(blue)} \qquad\qquad \text{(colorless)}$$

As shown in Eq. (56), when a column of immobilized dehydrogenase enzyme (DH_1) is perfused with a mixture of substrates (S) and nicotinamide adenine dinucleotide (NAD), reduced NAD($NADH_2$) and products (P) are produced. The concentration of $NADH_2$ in the column effluent stream is measured according to Eq. (57) as the decrease in the absorbance of an oxidized blue dye (Dye_{ox}, 2,6-dichlorophenolindophenol) at 620 mμ in the presence of a catalyst, phenazine methosulfate (PMS).

The detection of immobilized oxidase activity was based on the following reactions:

$$S + O_2 \xrightleftharpoons{OX_1} P + H_2O_2 \tag{58}$$

$$H_2O_2 + Dye_{red} \xrightarrow{peroxidase} H_2O + Dye_{ox} \tag{59}$$
$$\text{(colorless)} \qquad\qquad \text{(blue)}$$

According to Eq. (58), when a column of immobilized oxidase enzyme (OX_1) is perfused with a substrate (S) and O_2, products (P) and H_2O_2 are produced. The H_2O_2 in the column effluent stream is detected, as illustrated in Eq. (59), by measuring the increase in absorbance of an oxidized blue dye (Dye_{ox}, o-toluidine) at 620 mμ in the presence of a second enzyme, peroxidase.

Lyophilized acrylamide-entrapped glucose oxidase, when kept at 0° to 4°, showed no loss of activity after 3 months. Lyophilized acrylamide-entrapped lactic dehydrogenase lost about 10% of its activity per month over a period of 3 months. A hydrated column of glucose oxidase gel lost less than 5% of its activity in 6 weeks when kept in a refrigerator. A hydrated LDH column was less stable, losing most of its activity in about 3 months at 4°. Some attempts were made to determine whether or not the enzyme activity in the gel was actually more stable than in free solution. When an LDH-column was perfused continuously with lactate at 37°, no loss of activity was noticed for at least 10 hours. After 10 hours, the activity began to decrease; 50% of the activity was lost after about 20 hours had elapsed. The stability was a function of time, but not of flow rate, suggesting an actual denaturation process as opposed to the "wash out" of the enzyme. The same purified LDH enzyme used in the preparation of the gel was very unstable in solution when maintained at 37°, losing 90% of its activity in only 2 hours. After 2 hours, the activity remained constant at 10% of its original value.

An enzyme electrode is a miniature chemical transducer which functions by combining an electrochemical procedure with immobilized enzyme activity (Updike and Hicks, 1967). Two such enzyme electrodes have been described in the literature.

A glucose-specific electrode has been constructed by Updike and Hicks (1967) by polymerizing a gelatinous membrane of immobilized glucose oxidase over a polarographic oxygen electrode. The oxygen electrode measures the diffusion flow of oxygen through a plastic membrane. The current output of the electrode is a linear function of oxygen tension. Specificity for glucose is obtained by immobilizing the enzyme glucose oxidase in a layer of acrylamide gel, 25–50 μ thick, over the oxygen electrode as shown in Fig. 34. When the enzyme electrode is placed in contact with a biological solution or tissue, glucose and oxygen diffuse into the gel layer of immobilized enzyme. The diffusion flow of oxygen through the plastic membrane of the oxygen electrode is reduced in the presence of glucose oxidase and glucose by the enzyme-catalyzed reaction of glucose and oxygen [Eq. (60)].

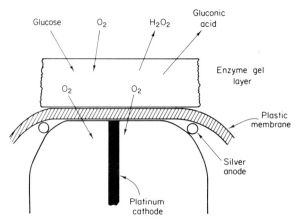

Fig. 34. Scheme of a glucose oxidase electrode. [Updike and Hicks (1967).]

$$\text{Glucose} + O_2 \xrightarrow{\text{glucose oxidase}} \text{gluconic acid} + H_2O_2 \quad (60)$$

When oxygen is in nonrate-limiting excess and the glucose concentration is well below the apparent Michaelis constant for the immobilized glucose oxidase, there is a linear relationship between glucose concentration and the decrease in oxygen tension. The output current of the enzyme electrode is measured after allowing sufficient time for the diffusion process to reach the steady state. This interval varies from about 30 seconds to 3 minutes, depending primarily on the thickness of the plastic and gel membranes. The glucose oxidase electrode referenced against a silver-silver chloride electrode could be used after appropriate standardization to determine the glucose content of whole blood and plasma. The glucose oxidase electrode is sensitive to temperature with an electrode output of about 5.3%/°C in the range 25°–40°, and can be made to function with sufficient sensitivity at 0°. The single cathode glucose oxidase electrode is sensitive to changes in the concentration of both glucose and oxygen. For *in vivo* monitoring of tissue glucose concentration a differential glucose oxidase electrode, independent of oxygen tension, could be constructed by using two glucose oxidase electrodes, one of which has been heat inactivated, referenced against a single silver-silver chloride electrode.

A urease electrode operating on a similar principle was obtained by polymerizing a membrane of acrylamide gel-immobilized enzyme over a cationic glass electrode which is responsive to ammonium ions (Gilbault and Montalvo, 1969). The layer of acrylamide gel-entrapped urease on the surface of the glass electrode was 60–350 μ thick. When the urease electrode is placed in contact with a urea solution, the substrate diffuses into the immobilized enzyme layer. The ammonium ion produced at the surface of the glass electrode by the enzyme-catalyzed decomposition of urea [Eq. (61)] is sensed by the ammonium-specific glass electrode which measures the activity of this monovalent cation in a manner analogous to pH determination with a glass electrode.

$$\text{Urea} + H_2O \xrightarrow{\text{urease}} 2NH_4^+ + CO_2 \qquad (61)$$

The potential of the ammonium electrode is measured after allowing sufficient time for the diffusion process to reach the steady state (25 to 60 seconds, depending on the thickness of the gel membrane). When the urea concentration is below the apparent Michaelis constant for the immobilized enzyme but above 0.6 mg of urea per 100 ml of solution, the potential of the electrode varies linearly with the logarithm of the urea concentration. The urease electrode has been applied for the determination of urea in body fluids (Guibault and Montalvo, 1970; Guilbault and Hrabankova, 1970).

The use of immobilized enzyme systems should extend the application of enzymes in analysis. Aside from the advantage of economy, another major advantage of the immobilized system, which should permit the development of many new analytical applications, is the fact that the products of the enzyme reaction are easily separated from the enzyme catalyst. Thus, an analytical method could be based on a series of enzymic reactions by utilizing a series of enzyme gel columns, even though the individual enzymes were not mutually compatible in a single solution. Furthermore, the exclusion of molecules from the enzyme reaction system trapped in the gels on the basis of molecular size in a manner similar to that for "gel filtration" techniques may help to reduce the number of interferences, which is a major problem with many enzymic methods of analysis.

Enzyme columns prepared by adsorbing a mold aminoacylase on DEAE-cellulose or DEAE-Sephadex have been utilized by Tosa et al. (1966, 1967a,b, 1969a,b) for the resolution of synthetic racemic amino acids. The DL-amino acids were acetylated and the N-acyl L-isomer specifically hydrolyzed by the enzyme. Under the appropriate conditions of flow rate and temperature, quantitative resolution of acetyl-DL-methionine into L-methionine and acetyl-D-methionine was attained (Tosa et al., 1966, 1967a). Continuously operated aminoacylase columns are at present being used in Japan for large scale preparation of optically active amino acids (Tosa et al., 1967b, 1969a,b). In batch processes in which crude enzyme extracts are employed, a purification step is necessary to isolate both antipodes in pure form. In the process based on the aminoacylase columns, contamination of products by foreign protein and colored substances originating from the crude enzyme extracts does not occur. The isolation of pure product was thus simplified concurrently with a considerable increase in overall yields. The aminoacylase columns retained 60% of their activity after continuous operation at 50° for 1 month. The loss of activity was most probably due to slow elution of the adsorbed enzyme. The activity of the column could, however, be regenerated by recharging with enzyme (Tosa et al., 1969b). It should be noted that at high substrate concentrations, a considerable increase in the rate of elution of adsorbed enzymes has been observed. This limitation can obviously be overcome by the covalent binding of the enzyme to an insoluble carrier.

Microcapsules made of thin spherical semipermeable nylon or collodion membranes, and comparable in size to tissue cells, have been used to encapsulate various enzymes (Chang, 1964, 1966; Chang et al., 1966). Such microcapsules could be made nonthrombogenic by an appropriate treatment. When these capsules were placed in a nonthrombogenic shunt chamber, blood could be circulated through the system for more than two hours without clotting (Chang, 1966; Chang et al., 1966). It is thus possible, in the case of enzyme deficiency diseases, to introduce the missing enzymes into the blood stream in microencapsulated form. Such a procedure might prevent antibody formation against the "foreign" enzyme and allergic complications on repeated treatment. A mixture of microencapsulated urease and ion-exchange resin, for example, might be of use in the removal of excess urea from the circulation under conditions of kidney failure.

References

Aldrich, F. L., Usdin, V. R., and Vasta, B. M., U.S. Army Report DA–18–108–405 CML–828. (1963).
Aldrich, F. L., Usdin, V. R., and Vasta, B. M. (1965). U.S. Pat. 3,223,593.

Alexander, B., Rimon, A., and Katchalski, E. (1965). *Fed. Proc., Fed. Amer. Soc. Exp. Biol.* **24**, 804.
Alexander, B., Rimon, A., and Katchalski, E., (1966). *Biochemistry* **5**, 792.
Arnold, W. N. (1966). *Arch. Biochem. Biophys.* **113**, 451.
Ashoor, S. H., Sair, R. A., Olson, N. F., and Richardson, T. (1971). *Biochim. Biophys. Acta* **229**, 423.
Atkinson, B., Swilley, E. L., Busch, A. W., and Williams, D. A. (1967). *Trans. Inst. Chem. Eng.* **45**, 257.
Axén, R., and Porath, J. (1966). *Nature (London)* **210**, 367.
Axén, R., Porath, J., and Ernback, S. 1967. *Nature (London)* **214**, 1302.
Axén, R., Heilbron, E., and Winter, A. (1969). *Biochim. Biophys. Acta* **191**, 478.
Axén, R., Myrin, P. A., and Janson, J. C. (1970). *Biopolymers*, **9**, 401.
Bachler, M. J., Strandberg, G. W., and Smiley, K. L. (1970). *Biotechnol. Bioeng.* **12**, 85
Baker, B. R. (1967). "Design of Active-Site-Directed Irreversible Enzyme Inhibitor," Chapter 13. Wiley, New York.
Bar-Eli, A., and Katchalski, E. (1960). *Nature (London)* **188**, 856.
Bar-Eli, A., and Katchalski, E. (1963). *J. Biol. Chem.* **238**, 1690.
Barker, S. A., Somers, P. J., and Epton, R. (1968). *Carbohyd. Res.* **8**, 491.
Barker, S. A., Somers, P. J., and Epton, R. (1969). *Carbohyd. Res.* **9**, 257.
Barker, S. A., Somers, P. J., Epton, R. and McLaren, J. V. (1970). *Carbohyd., Res.* **14**, 287.
Barnett, L. B., and Bull, H. B., (1959). *Biochim. Biophys. Acta* **36**, 244.
Bass, L., and McIlroy, D. K. (1968). *Biophys. J.* **8**, 99.
Bauman, E. K., Goodson, L. H., Guilbault, G. G., and Kramer, D. N. (1965). *Anal. Chem.* **37**, 1378.
Bauman, E. K., Goodson, L. H., and Thomson, J. R. (1967). *Anal. Biochem.* **19**, 587.
Bender, M. L., and Kézdy, F. J. (1965). *Annu. Rev. Biochem.* **34**, 49.
Bernfeld, P., and Wan, J. (1963). *Science* **142**, 678.
Bernfeld, P., Bieber, R. E., and MacDonnell, P. C. (1968). *Arch. Biochem. Biophys.* **127**, 779.
Bernfeld, P., Bieber, R. E., and Watson, D. M. (1969a). *Biochim. Biophys. Acta* **191**, 570.
Bernfeld, P., Bieber, R. E., Watson, D. M., and MacDonnell, P. C. (1969b). *Fed. Proc. Fed. Amer. Soc. Exp. Biol.* **28**, 534.
Blumberg, S., Schechter, I., and Berger, A. (1969). *Isr. J. Chem.* **7**, 125p.
Blumberg, S., Schechter, I., and Berger, A. (1970). *Eur. J. Biochem.* **15**, 97.
Blumenthal, R., Caplan, S. R., and Kedem O. (1967). *Biophys. J.* **7**, 735.
Brandenberger, H. (1956). *Rev. Ferment. Ind. Aliment.* **11**, 237.
Broun, G., Selegny, E., Avrameas, S., and Thomas, D. (1969). *Biochim. Biophys. Acta* **185**, 260.
Brown, H. D., Chattopadhyay, S. K., and Patel, A. (1966). *Biochem. Biophys. Res. Commun.* **25**, 304.
Brown, H. D., Chattopadhyay, S. K., and Patel, A. (1967). *Enzymologia* **32**, 205.
Campbell, D. H., Leuscher, E., and Lerman, L. S. (1951). *Proc. Nat. Acad. Sci. U.S.* **37**, 575.
Cebra, J. J., (1964). *J. Immunol.* **92**, 977.
Cebra, J. J., Givol, D., Silman, H. I., and Katchalski, E. (1961). *J. Biol. Chem.* **236**, 1720.
Cebra, J. J., Givol, D., and Katchalski, E. (1962). *J. Biol. Chem.* **237**, 751.
Chang, T. M. S. (1964). *Science* **146**, 524.

Chang, T. M. S. (1966). *Trans. Amer. Soc. Artif. Intern. Organs* **12**, 13.
Chang, T. M. S., McIntosh, F. C., and Mason, S. G. (1966). *Can. J. Physiol. Pharmacol.* **44**, 115.
Chibata, I., Tosa T., and Endo, N. (1966a). Jap. Pat. 7829.
Chibata, I., Tosa T., and Endo, N. (1966b). Jap. Pat. 22,380.
Cohen, C., Slayter, H., Goldstein, L., Kucera, J., and Hall, C. (1966). *J. Mol. Biol.* **22**, 385.
Craven, G. R., and Gupta, V. (1970). *Proc. Nat. Acad. Sci. U.S.* **67**, 1329.
Cresswell, P., and Sanderson, A. R. (1970). *Biochem. J.* **119**, 447.
Cuatrecasas, P., Wilchek, M., and Anfinsen, C. B. (1968). *Proc. Nat. Acad. Sci U.S.* **61**, 636.
Davis, R. V., Blanken, R. M., and Beagle, R. J. (1969). *Biochemistry* **8**, 2706.
Degani, Y., and Miron, T. (1970). *Biochim. Biophys. Acta* **212**, 362.
De Groot, S. R., and Mazur, P. (1963). In "Non Equilibrium Thermodynamic," p. 43. North-Holland Publ. Amsterdam.
Denburg, J., and DeLuca, M. (1970). *Proc. Nat. Acad. Sci. U.S.* **67**, 1057.
Dixon, M., and Webb, E. C. (1964). " Enzymes," (2nd ed.) Academic Press, New York.
Doscher, M. S., and Richards, F. M. (1963). *J. Biol. Chem.* **238**, 2399.
Energy Coupling in Electron Transport. (1967). *Fed. Proc., Fed. Amer. Soc. Exp. Biol.* **26**, 1339–1379.
Engel, A., and Alexander, B., (1966). *Biochemistry* **5**, 3590.
Engel, A., and Alexander, B. (1971). *J. Biol. Chem.* **246**, 1213
Epstein, C. J., and Anfinsen, C. B. (1962). *J. Biol. Chem.* **237**, 2175.
Erlanger, B. F., Isambert, M. F., and Michelson, A. M. (1970). *Biochem. Biophys. Res. Comm.* **40**, 70.
Fraenkel–Conrat, H. (1959). *In* " The Enzymes" (P. D. Boyer, H. Lardy, and K. Myrbäck, eds.), 2nd rev. ed., Vol. 1, p. 589. Academic Press, New York.
Fritz, H., Schult, H., Neudecker, M., and Werle, E. (1966). *Angew. Chem.* **78**, 775; (1966). *Angew. Chem., Int. Ed. Engl.* **5**, 735.
Fritz, H., Schult, H., Hutzel, M., Wiedermann, M., and Werle, E. (1967). *Hoppe-Seyler's Z. Physiol. Chem.* **348**, 308.
Fritz, H., Hochstrasser, K., Werle, E., Brey, E., and Gebhardt, B. M. (1968). *Z. Anal. Chem.* **243**, 452.
Fritz, H., Gebhardt, B. M., Fink, E., Schramm, W., and Werle, E. (1969). *Hoppe-Seyler's Z. Physiol. Chem.* **305**, 129.
Gabel, D., and Hofsten, B. v. (1970). *Eur. J. Biochem.* **15**, 410.
Gabel, D., Vretblad, P., Axén, R., and Porath, J. (1970). *Biochim. Biophys. Acta* **214**, 561.
Garen, A., and Levinthal, C. (1960). *Biochim. Biophys. Acta* **38**, 470.
Ginzburg, B. Z., and Katchalsky, A. (1963). *J. Gen. Physiol.* **47**, 403.
Givol, D., Weinstein, Y., Gorecki, M., and Wilchek, M. (1970). *Biochem. Biophys. Res. Commun.* **38**, 825.
Glassmeyer, C. K., and Ogle, J. D. (1971). *Biochemistry* **10**, 786.
Goldfeld, M. G., Vorobeva, E. S. and Poltorak, O. M. (1966). *Zh. Fiz. Khim.* **40**, 2594.
Goldman, R., and Katchalski, E. (1969). *3rd Int. Biophys. Congr. Int. Union Pure Appl. Biophys. 1969* p. 59.
Goldman, R., and Lenhoff, H., (1969). Unpublished data.
Goldman, R., Silman, H. I., Caplan, S. R., Kedem, O., and Katchalski, E. (1965). *Science* **150**, 758.

Goldman, R., Kedem, O., Silman, I. H., Caplan, S. R., and Katchalski, E. (1968a). *Biochemistry* **7**, 486.
Goldman, R., Kedem, O., and Katchalski, E. (1968b). *Biochemistry* **7**, 4518.
Goldman, R., Kedem, O., and Katchalski, E. (1971). *Biochemistry* **10**, 165.
Goldman, R., and Katchalski, E. (1971). *J. Theor. Biol.* **32** (in press).
Goldstein, L. (1969). *In* "Fermentation Advances" (D. Perlman, ed.), p. 391. Academic Press, New York.
Goldstein, L. (1970). *Methods Enzymol.* **19**, 935.
Goldstein, L. (1971). Manuscript in preparation.
Goldstein, L., and Katchalski, E.,(1968). *Z. Anal. Chem.* **243**, 375.
Goldstein, L., Levin, Y., and Katchalski, E. (1964). *Biochemistry* **3**, 1913.
Goldstein, L., Pecht, M., Blumberg, S., Atlas, D., and Levin, Y. (1970). *Biochemistry* **19**, 2322.
Green, M. L., and Crutchfield, G. (1969). *Biochem. J.* **115**, 183.
Gregor, H. P., and Sollner, K. (1946). *J. Phys. Chem.* **50**, 53.
Grubhofer, H., and Schleith, L. Z. (1953). *Naturwissenschafkn.* **40**, 508.
Grubhofer, H., and Schleith, L. Z. (1954). *Hoppe-Seyler's Z. Physiol. Chem.* **297**, 108.
Guilbault, G. G., and Kramer, D. N. (1965) *Anal. Chem.* **37**, 1675.
Guilbault, G. G., and Montalvo, J. G. (1969). *J. Am. Chem. Soc.* **91**, 2164.
Guilbault, G. G., and Das, J. (1970). *Analyt. Biochem.* **33**, 341.
Guilbault, G. G., and Hrabankova, E. (1970). *Anal. Chim. Acta* **52**, 287.
Guilbault, G. G., and Montalvo, J. G. (1970). *J. Am. Chem. Soc.* **92**, 2533.
Gurvich, A. E. (1957). *Biochemistry (USSR)* **22**, 977.
Gutfreund, H., (1955). *Trans. Faraday Soc.* **51**, 441.
Gutman, M., and Rimon, A. (1964). *Can. J. Biochem.* **42**, 1339.
Habeeb, A. F. S. A. (1967). *Arch. Biochem. Biophys.* **119**, 264.
Hartman, F. C., and Wold, F. (1967). *Biochemistry* **6**, 2439.
Haynes, R., and Walsh, K. A. (1969). *Biochem. Biophys. Res. Commun.* **36**, 235.
Heinz, E. (1967). *Annu. Rev. Physiol.* **29**, 21.
Helfferich, F. (1962a). *In* "Ion Exchange," p. 253. McGraw-Hill, New York.
Helfferich, F. (1962b). *In* "Ion Exchange," p. 519. McGraw–Hill, New York.
Herzig, D. J., Rees, A. W., and Day, R. A. (1964). *Biopolymers* **2**, 349.
Hicks, G. P., and Updike, S. J. (1966). *Anal. Chem.* **38**, 726.
Hoave, D. G., and Koshland, D. E., Jr. (1967). *J. Biol. Chem.* **242**, 2447.
Hoffer, E., and Kedem, O. (1968). *Desalination* **5**, 167.
Hornby, W. E., Lilly, M. D., and Crook, E. M. (1966). *Biochem. J.* **98**, 420.
Hornby, W. E., Lilly, M. D., and Crook, E. M. (1968). *Biochem. J.* **107**, 669.
Hornby, W. E., Filippuson, H., and McDonald, A. (1970). *FEBS Letters* **9**, 8.
Hornby, W. E., and Filippuson, H. (1970). *Biochim. Biophys. Acta* **220**, 343.
Hummel, J. P., and Anderson, B. S. (1965). *Arch. Biochem. Biophys.* **112**, 443.
Hussain, Q. Z., and Newcomb, T. F. (1964). *Proc. Soc. Exp. Biol. Med.* **115**, 301.
Inman, J. K., and Dintzis. H. M. (1969). *Biochemistry* **8**, 4074.
Jansen, E. F., and Olson, A. C. (1969) *Arch. Biochem. Biophys.* **129**, 221.
Jaquet, H., and Cebra J. J. (1965). *Biochemistry* **4**, 954.
Johnson, M. J. (1967). *J. Bacteriol.* **94**, 101.
Kallos, J. (1964). *Biochim. Biophys. Acta* **89**, 364.
Katchalski, E. (1962). *In* "Polyamino Acids, Polypeptides, Proteins," *Proc. Int. Symp., 1st, 1961,* (M. A. Stahmann, ed.), p. 283. Univ. Wisconsin Press.
Kay, G., and Crook, E. M. (1967). *Nature (London)* **216**, 514.

Kay, G., Lilly, M. D., Sharp, A. K., and Wilson, R. J. H. (1968). *Nature (London)* **217**, 641.
Kay, G., and Lilly, M. D. (1970). *Biochim. Biophys. Acta* **198**, 276.
Kirimura, J., and Yoshida, T. (1966). U.S. Pat. 3,243,356.
Kobamoto, N., Löfroth, G., Camp, P., Van Amburg, G., and Augenstein, L. (1966). *Biochem. Biophys. Res. Commun.* **24**, 622.
Kominz, D. R., Mitchell, E. R., Nihei, T., and Kay, C. M. (1965). *Biochemistry* **4**, 2373.
Leuschner, F. (1964). Brit. Pat. 953,414.
Leuschner, F. (1966). Ger. Pat, 1,227,855.
Levin, Y., Pecht, M., Goldstein, L., and Katchalski, E. (1964). *Biochemistry* **3**, 1905.
Lewis, M., and Sollner, K. (1959). *J. Electrochem. Soc.* **106**, 347.
Lilly, M. D., and Sharp, A. K. (1968). *Chem. Eng. (London)*. CE12.
Lilly, M. D., Money, C., Hornby, W. E., and Crook, E. M. (1965). *Biochem. J.* **95**, 45p.
Lilly, M. D., Hornby, W. E., and Crook, E. M., (1966). *Biochem. J.* **100**, 718.
Lilly, M. D., Kay, G., Sharp, A. K., and Wilson, R. J. H. (1968). *Biochem. J.* **107**, 5p.
Lowey, S. (1968). *Symp. Fibrous Proteins, 1967* p. 124. Plenum Press, New York.
Lowey, S., Goldstein, L., and Luck, S. (1966). *Biochem. Z.* **345**, 248.
Lowey, S., Goldstein, L., Cohen, C., and Luck, S. M. (1967). *J. Mol. Biol.* **23**, 287.
Lowey, S., Slayter, H. S., Weeds, G., and Baker, H. (1968). *J. Mol. Biol.* **42**, 1.
Lynn, J., and Falb, R. D. (1969). *Abstr. 1158th Meet. Amer. Chem. Soc.* Contrib. 298, Biol. Sect.
McLaren, A. D. (1957). *Science* **125**, 697.
McLaren, A. D. (1960). *Enzymologia* **21**, 356.
McLaren, A. D., and Babcock, K. L. (1959). *In* "Subcellular Particles" (H. Hayashi, ed.), p. 23. Ronald Press, New York.
McLaren, A. D., and Estermann, E. F. (1956). *Arch. Biochem. Biophys.* **61**, 158.
McLaren, A. D., and Estermann, E. F. (1957). *Arch. Biochem. Biophys.* **68**, 157.
McLaren, A. D., Peterson, G. H., and Barshad, I. (1958). *Soil Sci. Soc. Amer., Proc.* **22**, 239.
Manecke, G. (1962). *Pure Appl. Chem.* **4**, 507.
Manecke, G. (1964). *Naturwissenschaften* **51**, 25.
Manecke, G., and Förster, H. J. (1966). *Makromol. Chem.* **91**, 136.
Manecke, G., and Günzel, G. (1967). *Naturwissenschaften* **54**, 647.
Manecke, G., and Singer, S. (1960). *Makromol. Chem.* **39**, 13.
Manfrey, P. S., and King, M. V. (1965). *Biochim. Biophys. Acta* **105**, 178.
Micheel, F., and Evers, J. (1949). *Makromol. Chem.* **3**, 200.
Mitchell, P. (1966). *Biol. Rev.* **41**, 445.
Mitchell, P. (1967). *Advan. Enzymol.* **29**, 33.
Mitz, M. A. (1956). *Science* **123**, 1076.
Mitz, M. A., and Schleuter, R. J. (1959). *J. Amer. Chem. Soc.* **81** 4024.
Mitz, M. A., and Summaria, L. J. (1961). *Nature (London)* **189**, 576.
Mitz, M. A., and Yanari, S. S. (1964). U.S. Pat. 3,126,324.
Mosbach, K., and Mosbach, R. (1966). *Acta Chem. Scand.* **20**, 2807.
Mosbach, K., and Larsson, P. O. (1970). *Biotechnol. Bioengin.* **12**, 19.
Mosbach, K., and Mattiasson, B. (1970). *Acta Chem. Scand.* **24**, 2093.
Mosbach, K. (1970). *Acta Chem. Scand.* **24**, 2084.
Nernst, W. Z. (1904). *Z. Phys. Chem.* **47**, 52.
Newcomb, T. F., and Hoshida, M. (1965) *Scand. J. Clin. Lab. Invest.* **17**., Suppl. **84**, 61.
Nikolayev, A. Y. (1962). *Biokhimiya* **27**, 843.
Nikolayev, A. Y., and Mardashev, S. R. (1961). *Biokhimiya* **26**, 641; see *Biochemistry (USSR)* **26**, 565 (1961).
Nisonoff, A., Wissler, F. C., and Lipman, L. N. (1960a). *Science* **132**, 1770.

Nisonoff, A., Wissler, F. C., Lipman, L. N., and Woernley, D. L. (1960b). *Arch. Biochem. Biophys.* **89**, 230.
Ogata, K., Ottesen, M., and Svendsen, I. (1968). *Biochim. Biophys. Acta* **159**, 403.
Ong, E. B., Tsang, Y., and Perlmann, G. E. (1966). *J. Biol. Chem.* **241**, 5661.
Ottesen, M., and Svensson, B. (1971). *Compt. Rend. Trav. Lab. Carlsberg* **38**, 171.
Patchornik, A. (1962). Isr. Pat. 18,207.
Patel, A. B., Pennington, S. N., and Brown, H. D. (1969). *Biochim. Biophys. Acta* **178**, 26.
Patel, R. P., and Price, S. (1961). *Biopolymers* **5**, 583.
Patel, R. P., Lopiekes, D. V., Brown, S. R., and Price, S. (1967). *Biopolymers* **5**, 577.
Pennington, S. N., Brown, H. D., Patel, A. B., and Knowles, C. O. (1968). *Biochim. Biophys. Acta* **167**, 479.
Penzer, G. R., and Radda G. K. (1967). *Nature (London)* **213**, 251.
Poltorak, O. M., and Vorobeva, E. S. (1966). *Zh. Fiz. Khim.* **40**, 1665.
Porath, J. (1967). *In* "Gamma Globulins" (J. Killander, ed.), Nobel Symp. No. 3, p. 287. Wiley (Interscience), New York.
Porath, J., Axén, R., and Ernback, S. (1967). *Nature (London)* **215**, 1491.
Porter, R. R. (1959). *Biochem. J.* **73**, 119.
Quiocho, F. A., and Richards, F. M. (1964). *Proc. Nat. Acad. Sci. U.S.* **52**, 833.
Quiocho, F. A., and Richards, F. M. (1966a). *Biochemistry* **5**, 4062.
Quiocho, F. A., and Richards, F. M. (1966b) *Biochemistry* **5**, 4077.
Richards, F. M. (1963). *Annu. Rev. Biochem.* **32**, 268.
Riesel, E., and Katchalski, E. (1964). *J. Biol. Chem.* **239**, 1521.
Rimon, A., Gutman, M., and Rimon S. (1963). *Biochim. Biophys. Acta* **73**, 301.
Rimon, S., Stupp, Y., and Rimon, A. (1966). *Can. J. Biochem.* **44**, 415.
Schejter, A., and Bar-Eli, A. (1970). *Arch. Biochem. Biophys.* **136**, 325.
Schick, H. F., and Singer, S. J. (1961). *J. Biol. Chem.* **236**, 2447.
Sehon, A. H. (1967). *Symp. Ser. Immunobiol. Stand.* **4**, 51.
Sélégny, E., Avrameas, S., Broun, G., and Thomas, D. (1968). *Compt. rend. Acad. Sci., Ser. C* **266**, 1431.
Seki, T., Jenssen, T. A., Levin, Y., and Erdös, E. G. (1970a). *Nature* **225**, 864.
Seki, T., Yang, H. Y. T., Levin, Y., Jenssen, T. A., and Erdös, E. G. (1970b). *In* "Bradykinin and Related Kinins: Cardiovascular, Biochemical and Neural Actions," p. 23. Plenum Press, New York.
Shaltiel, Sh., Mizrahi, R. Stupp, Y., and Sela, M (1970). *Europ. J. Biochem.* **14**, 509.
Sharp, A. K., Kay, G., and Lilly, M. D. (1969). *Biotechnol. Bioengin.* **11**, 363.
Siegler, P. B., Blow, D. M., Matthews, B. W., and Henderson, R. (1968). *J. Mol. Biol.* **35**, 143.
Silman, I. H., and Karlin, A. (1967). *Proc. Nat. Acad. Sci. U.S.* **58**, 1664.
Silman, I. H., and Katchalski, E. (1966). *Annu. Rev. Biochem.* **35**, 873.
Silman, I. H., Wellner, D., and Katchalski, E. (1963). *Isr. J. Chem.* **1**, 65.
Silman, I. H., Albu-Weissenberg, M., and Katchalski, E. (1966). *Biopolymers* **4**. 441.
Slayter, H. S., and Lowey, S. (1967). *Proc. Nat. Acad. Sci. U.S.* **58**, 1611.
Sluyterman, L. A. AE., and De Graaf, M. J. M. (1969). *Biochim. Biophys. Acta* **171**, 277.
Sri Ram, J., Terminiello, L., Bier, M., and Nord, F. F. (1954). *Arch. Biohcem. Biophys.* **52**, 464.
Sri Ram, J., Bier, M., and Maurer, P. H., (1962). *Advan. Enzymol.* **24**, 105.
Steinbuch, M., and Pejaudier, L. (1964). *Bibl. Haematol. (Pavia)* **19**, 169.
Stone, I. (1955). U.S. Pat. 2,717,852.
Sundaram, P. V., Tweedale, A., and Laidler, K. J. (1970). *Canad. J. Chem.* **48**, 1498.

Sundaram, P. V., and Hornby, W. E. (1970). *FEBS Letters* **10**, 325.
Surinov, B. P., and Manoylov, S. E. (1966), *Biokhimiya* **31**, 387.
Suzuki, H., Ozawa, Y., and Maeda, H. (1966). *Agr. Biol. Chem.* **30**, 807.
Tanford, C. (1961). "Physical Chemistry of Macromolecules." Wiley, New York.
Theorell, H., Chance, B., and Yonetani, T. (1966). *J. Mol. Biol.* **17**, 513.
Thiele, E. W. (1939). *Ind. Eng. Chem.* **31**, 916.
Tosa, T., Mori, T., Fuse, N., and Chibata, I. (1966). *Enzymologia* **31**, 214 and 225.
Tosa, T., Mori, T., Fuse, N., and Chibata, I. (1967a). *Enzymologia* **32**, 153.
Tosa, T., Mori, T., Fuse, N., and Chibata, I. (1967b). *Biotechnol. Bioeng.* **9**, 603.
Tosa, T., Mori, T., Fuse, N., and Chibata, I. (1969a). *Agr. Biol. Chem.* **33**, 1047.
Tosa, T., Mori, T., and Chibata, I. (1969b). *Agr. Biol. Chem.* **33**, 1053.
Trentham, D. R., and Gutfreund, H. (1968). *Biochem. J.* **106**, 455.
Updike, S. J., and Hicks, G. P. (1967). *Nature (London)* **214**, 986.
Usami, A., and Taketomi, N. (1965). *J. Ferment. Technol.* **23**, 267.
Vallee, B. L., and Riordan, J. F. (1969) *Annu. Rev. Biochem.* **38**, 733.
van Duijn, P., Pascoe, E., and van der Ploeg, M. (1967). *J. Histochem. Cytochem.* **15**, 631.
Vorobeva, E. S., and Poltorak, O. M. (1966a). *Zh. Fiz. Khim.* **40**, 2596.
Vorobeva, E. S., and Poltorak, O. M. (1966b). *Vestn. Mosk. Univ., Khim.* **21**, 17.
Wagner, T., Hsu, C. J., and Kelleher, G. (1968). *Biochem. J.* **108**, 892.
Weetall, H. H. (1969a). *Science* **166**, 615.
Weetall, H. H. (1969b) *Abstr. 158th Meet. Amer. Chem. Soc.* Contrib. 153, Biol. Sect.
Weetall, H. H., and Weliky, N. (1966). *Anal. Biochem.* **14**. 160.
Weetall, H. H., and Hersh, L. S. (1970). *Biochim. Biophys. Acta* **256**, 54.
Weetall, H. H. (1969). *Nature* **223**, 959.
Weetall, H. H., and Hersh, L. S. (1969). *Biochim. Biophys. Acta* **185**, 464.
Weliky, N., and Weetall, H. H. (1965). *Immunochemistry* **2**, 293.
Weliky, N., Brown, F. S., and Dale, E. C. (1969). *Arch. Biochem. Biophys.* **131**, 1.
Westman, T. L. (1969). *Biochem. Biophys. Res. Commun.* **35**, 313.
Wharton, C. W., Crook, E. M., and Brocklehurst, K. (1968a). *Eur. J. Biochem.* **6**, 565.
Wharton, C. W., Crook, E. M., and Brocklehurst, K. (1968b). *Eur. J. Biochem.* **6**, 572.
Wheeler, A. (1951). *Advan. Catal. Relat. Subj.* **3**, 249.
Wheeler, K. P., Edwards, B. A., and Whittam, R. (1969). *Biochim. Biophys. Acta* **191**, 187.
Whitaker, J. R., and Bender, M. L. (1965). *J. Amer. Chem. Soc.* **87**, 2728.
Whittam, R., Edwards, B. A., and Wheeler, K. P. (1968). *Biochem. J.* **107**, 3p.
Wieland, T., Determann, H., and Bünnig, K. (1966). *Z. Naturforsch. B* **21**, 1003.
Wilchek, M. (1969). *Isr. J. Chem.* **7**, 124p.
Wilson, R. J. H., Kay, G., and Lilly, M. D. (1968a). *Biochem. J.* **108**, 845.
Wilson, R. J. H., Kay, G., and Lilly, M. D. (1968b). *Biochem. J.* **109**, 137.
Wilson, R. J. H., and Lilly, M. D. (1969). *Biotechnol. Bioengin.* **11**, 349.
Wold, F., (1961). *J. Biol. Chem.* **236**, 106.
Zahn, H. and Meienhofer, H. (1958). *Makromol. Chem.* **26**, 126 and 153.
Zahn, H., Growitz, F., and von Heyl, G. C. (1962). *Kolloid-Z.* **180**, 26.
Zingaro, R. A., and Uziel, M. (1970). *Biochim. Biophys. Acta* **213**, 371.
Zittle, C. A. (1953). *Advan. Enzymol.* **44**, 319.

CHAPTER 2 Selective Adsorbents Based on Biochemical Specificity

PEDRO CUATRECASAS

I.	Introduction	79
II.	Water-Insoluble Carriers	82
III.	Selection of the Ligand	84
IV.	Covalent Linking Reactions	86
	A. Cellulose and Polystyrene Supports	86
	B. Beaded Agarose Supports	86
	C. Polyacrylamide Supports	91
V.	Conditions for Chromatography on Affinity Columns	94
VI.	Importance of Anchoring Arms	95
VII.	Specific Adsorbents for Protein Purification	97
VIII.	Specific Nucleic Acid Adsorbents	104
IX.	Isolation of Cells, Receptor Structures, and Other Particulate Cell Structures	105
	References	107

I. INTRODUCTION

Considerable interest has recently been directed to the preparation of water-insoluble derivatives useful in the isolation and purification of biologically active macromolecules. This approach, which has recently been referred to as "affinity chromatography" (Cuatrecasas, 1970b,c, 1971a,b; Cuatrecasas and Anfinsen 1971a,b; Cuatrecasas *et al.*, 1968; Jerina and Cuatrecasas, 1970), is based on the unique biological property of many proteins or polypeptides to bind ligands specifically and reversibly. Unlike the conventional procedures utilized for protein purification which are based on physicochemical criteria, selective adsorbents exploit the specific functional properties of macromolecules. This type of "functional purification" is related in principle to the use of "immunoadsorbents," which were first characterized and developed by Campbell *et al.* (1951) for the chromatographic separation of antibodies, and have since been employed widely as standard immunochemical procedures (Silman and Katchalski, 1966).

A selective adsorbent is prepared by attaching a specific competitive inhibitor, or other ligand, to a carefully selected insoluble polymer or gel. Although various noncovalent ways of attaching ligands to insoluble carriers are theoretically feasible (e.g., by physical adsorption), covalent attachment will be considered here since up to now this has been the only procedure that is generally useful. Furthermore, there is the serious risk in using noncovalent procedures that some of the ligand may desorb during the chromatographic experiment. Making a ligand insoluble by polymerizing it through intermolecular cross-links is another alternative which has not yet been well explored. Such polymerization, however, is likely to adversely alter the affinity of the protein toward the constituent ligand, and the physical properties of polymers prepared in this way are likely to be unsuitable for chromatographic experiments.

Purification is accomplished when a crude solution containing the protein or other macromolecule is passed through a column of the selective adsorbent. Those molecules not exhibiting appreciable affinity for the ligand will pass unretarded through the column, and those which recognize the ligand will be retarded to an extent determined by the affinity constant of the experimental conditions. If the effective binding is weak, the specific macromolecule will emerge near the breakthrough volume only slightly behind or superimposed on the major protein component. On the other hand, strong binding will cause the macromolecule to be adsorbed to the upper part of the column, where it will resist elution even after extensive washing with the standard buffer. In these cases it is necessary to change the buffer in order to elute the specifically adsorbed protein. The pH or the ionic strength, for example, can be changed to concentrations known to cause dissociation of the interacting species in solution. Alternatively, elution can be effected by adding protein denaturants (urea, guanidine, detergents), competitive inhibitors, allosteric ligands, or substrates to the buffer. In some cases, to be described later, effective elution can be achieved only by cleaving the intact ligand-macromolecular complex from the matrix.

Successful application of these methods will depend in large part on how closely the chosen experimental conditions permit the ligand-protein interaction to simulate the reactions observed when the components are free in solution. Hence, careful consideration must be given to the nature of the solid carrier, stereochemical alteration of the ligand resulting from its covalent linkage to the carrier, the distance separating the ligand from the backbone of solid matrix, the steric properties and flexibility of the group interposed between the ligand and the backbone, the stability of chemical bonds, and the experimental conditions selected for adsorption, washing, and elution. The specific conditions selected for purification of a particular macromolecule are likely be highly individual since they will reflect the specific properties of selective interaction with a particular ligand.

Affinity chromatography can, in principle, be applied to a wide variety of macromolecular-ligand systems. For example, as summarized in this article, specific adsorbents may be prepared to purify enzymes, antibodies, antigens, nucleic acids, vitamin binding proteins, transport proteins, repressor proteins, drug or hormone receptors, sulfhydryl-containing proteins, peptides formed by organic synthesis, and intact cell populations.

Affinity chromatography may also be useful in concentrating dilute solutions of protein, in removing denatured forms of a purified protein, and in the separation and resolution of protein components, resulting from specific chemical modifications of purified proteins (Cuatrecasas 1970b,c; Cuatrecasas and Anfinsen, 1971a,b; Cuatrecasas et al., 1968; Jerina and Cuatrecasas, 1970). Other useful applications of water-insoluble adsorbents, such as their utilization as primers for the synthesis of specific nucleic acids (Jovin and Kornberg, 1968), will be presented in later sections.

The inherent advantages of this method of purification are the rapidity and ease of the procedure, the rapid separation of the protein to be purified from inhibitors and destructive contaminants, e.g., proteases, protection from denaturation during purification by stabilizing the protein structure by ligand binding at the active site, the possibility of rapidly separating the soluble and solid components by centrifugation or filtration, and the frequent reutilization of the same adsorbent.

The enormous potential utility to many fields of biochemistry and medicine of the principles and procedures described in this article is yet to be realized. It must be emphasized that properly prepared and characterized water-insoluble ligands should prove of great value in basic studies exploring the nature of ligand-macromolecule interactions. It is generally accepted that insolubilized enzymes are important model systems for studying the behavior of enzymes which exist *in vivo* in membrane complexes. Similarly, it is likely that many enzymic processes, such as those catalyzed by multienzyme or functionally coupled enzyme systems, are the results of interactions with substrates, regulators, or inhibitors not truly free in solution but directed or immobilized in specific ways. Water-insoluble initiators, primers, or templates can also be used in unique ways to examine the behavior of some synthetic enzymes. This use will be illustrated later with studies of polymerases and insoluble polynucleotides (Jovin and Kornberg, 1968). Another example, which will be described in a later section, of the value of these techniques as basic biochemical tools is the study of the binding of free ligands to insolubilized enzyme subunits which would ordinarily aggregate in solution in the absence of a protein denaturant (Hennig and Ginsburg, 1970). The study of the biological activity of insolubilized derivatives of peptides and proteins which exist in solution in subunit form, or which display considerable propensity for aggregation, may also shed some light on the basic structural unit required for biologic activity. This point will be illustrated later with studies of insolubilized derivatives of insulin (Cuatrecasas, 1969a).

II. WATER-INSOLUBLE CARRIERS

Until recently virtually all water-insoluble carriers have been derivatives of cellulose or polystyrene and, with few exceptions, only enzymes or antigens have been covalently attached to these supports. Excellent reviews covering the preparation and use of such derivatives are available (Sehon, 1962, 1963; Manecke, 1962; Katchalski, 1962; Porter and Press, 1962; Franklin, 1964; Silman and Katchalski, 1966), and a separate chapter in this book reviews in detail various procedures available for rendering enzymes insoluble. The major efforts here will be devoted to considering the preparation of insoluble ligands, to describing the newer carriers (agarose, polyacrylamide) which have proved or promise to be superior to the previously used carriers (cellulose, polystyrene), and to discussing the use of these insoluble ligand derivatives in purifying macromolecules by affinity chromatography.

In general, the conditions under which effective adsorbents for enzyme purification are prepared are much more stringent than those required to prepare immunoadsorbents (Sehon, 1962; Silman and Katchalski, 1966; Cuatrecasas, 1969a,b). Protein antigens generally possess many antigenic determinants, and there are many functional groups on the surface of the protein which can serve as attachment links between it and the solid support. Hence, it is likely that random, imprecise, or poorly controlled attachment of the protein to insoluble carriers will result in a mixed population of molecules which are linked by groups located in different regions. A variety of antigenic determinants will therefore be exposed to the solvent. Conversely, since the manner in which a ligand can be covalently attached to a solid support without destroying or severely compromising its capacity to be strongly recognized by an enzyme active site is quite limited, the chemical procedures for the attachment of ligands must be very carefully controlled and defined. Another feature of most antibody-antigen reactions which contributes to the ease of preparation of immunoadsorbents compared to enzyme-specific adsorbents is the very high affinity of these kinds of interactions compared with most enzyme-substrate or inhibitor interactions, while the dissociation constants of antibody-antigen interactions are generally smaller than 10^{-7} M. A third general factor to be considered in comparing the problems involved in the preparation of immunoadsorbents and of enzyme-specific adsorbents is the steric restriction which may be imposed by the solid support. A globular protein antigen attached to an insoluble carrier will have antigenic determinants which are located at some distance from the backbone of the matrix and, therefore, are likely to be exposed to the solvent. In contrast, a small ligand attached directly to some carrier will be closer to the polymer backbone. The polymer may interfere physically with the proper interaction of the components, whether they are soluble or not in water. The importance of attaching ligands at a good distance from the matrix backbone (Cuatrecasas, 1970b,c) will be discussed and illustrated in a later section.

Some of the above considerations probably explain in part why until recently the purification of enzymes by affinity chromatography has met with scant success. The nature of the insoluble support or carrier used is of considerable importance. It will be discussed now in some detail.

An ideal insoluble carrier for affinity chromatography of enzymes should possess the following properties: it must interact very weakly with proteins in general, to minimize the nonspecific adsorption of proteins; it should exhibit good flow properties which are retained after coupling; it must possess chemical groups which can be activated or modified under conditions innocuous to the structure of the matrix to allow the chemical linkage of a variety of ligands (these chemical groups should be abundant in order to allow attainment of a high effective concentration of coupled inhibitor, i.e., capacity, so that satisfactory adsorption can be obtained even with protein-inhibitor systems of low affinity); it must be mechanically and chemically stable to the conditions of coupling and to the various conditions of pH, ionic strength, temperature, and presence of denaturants such as urea, guanidine hydrochloride, and detergents which may be needed for adsorption or elution because such properties also permit repeated use of the specific adsorbent; and it should form a very loose, porous network which permits uniform and unimpaired entry and exit of large macromolecules throughout the entire matrix.

In addition, the gel particles should preferably be uniform, spherical, and rigid. A high degree of porosity is an important consideration of ligand-protein systems of relatively weak affinity (dissociation constant of 10^{-4} M or greater), since the concentration of ligand freely *available* to the protein must be quite high to permit interactions strong enough to physically retard the downward migration of the protein through the column. These restrictions probably explain why certain solid supports such as cellulose and polystyrene, which have been used so successfully as immunoadsorbents, have been of only minimal value in the purification of enzymes.

In some cases cellulose derivatives have been used in purifying enzymes (Lerman, 1963; Arsenis and McCormick, 1964, 1966) and avidin (McCormick, 1965). Derivatives of cellulose also have been used to purify nucleotides (Sander *et al.*, 1966), complementary strands of nucleic acids (Bautz and Holt, 1962; Adler and Rich, 1962; Gilham, 1962; Gilham and Robinson, 1964), and certain species of transfer RNA (Erham *et al.*, 1965). Polynucleotide derivatives of cellulose have also been used successfully to study enzymes catalyzing nucleic acid reactions (Cozzarelli *et al.*, 1967; Jovin and Kornberg, 1968). Although derivatives of cellulose may be useful in some instances, they are generally less useful than the agarose derivatives for enzyme purification because their fibrous and nonuniform character impedes proper penetration of large protein molecules.

Highly hydrophobic polymers such as polystyrene display poor communication between the aqueous and solid phases, and there is considerable nonspecific adsorption of protein to these supports. A technique for the immobilization of organic substances on glass surfaces, described recently by Weetall and Hersh (1969), may be a promising tool in preparing specific adsorbents.

Various hydrophilic polymers derived from polysaccharide have recently been found to be very useful insoluble carriers. Commercially available, cross-linked dextran derivatives (Sephadex) possess most of the desirable features listed above except for their low porosity. Hence they are relatively ineffective as adsorbents for enzyme purification of even low molecular weight, although they have been used successfully in preparing immunoadsorbents (Wide et al., 1967). However, beaded derivatives of agarose (Hjerten, 1962), another polysaccharide polymer, have nearly all the properties of an ideal adsorbent (Cuatrecasas, 1971b,c; Cuatrecasas and Anfinsen, 1971a,b; Cuatrecasas et al., 1968; Jerina and Cuatrecasas, 1970) and are commerically available as Sepharose. As the beaded agarose derivatives are very loose structured, molecules with a molecular weight in the millions diffuse readily through the matrix. These cross-linked polysaccharides readily undergo substitution reactions when activated by cyanogen halides (Porath et al., 1967; Axén et al., 1967), are very stable, and have a moderately high capacity for substitution. Synthetic polyacrylamide gels also possess many desirable features and are available commerically as spherical beads in pregraded sizes and porosities. Recently described derivatization procedures (Cuatrecasas, 1970b,c; Inman and Dintzis, 1969) permit the attachment of a variety of ligands and proteins to polyacrylamide beads. These beads have uniform physical properties and porosity, and the polyethylene backbone gives them physical and chemical stability. Preformed beads are available which permit penetration of proteins with molecular weights of about one half million, but their porosity is diminished during the chemical modifications required for the attachment of ligands. In this respect the polyacrylamide beads are inferior to those of agarose (Cuatrecasas, 1970b). The principal advantage of polyacrylamide is its very large number of modifiable groups (carboxamide). Thus, highly substituted derivatives may be prepared for use in purifying enzymes which are not strongly attached to the ligand. If highly porous acrylamide beads become widely available they will be nearly ideal supports for enzymes affinity chromatography.

III. SELECTION OF THE LIGAND

In preparing selective enzyme adsorbents the small molecule to be covalently linked to the solid support must display special affinity for the macromolecule to be purified. It can be a substrate analog, an effector, a cofactor, a vitamin, and in special cases, a substrate. Enzymes requiring two substrates for reaction may be approached by immobilizing one of the substrates, provided that one substrate is sufficiently well attached in the absence of the other. Also, a substrate may be used if it binds to the enzyme under some conditions which do not favor catalysis, i.e., in the absence of metal ion, at low temperatures, of if the pH dependence of K_m and of k_{cat} are different.

The small molecule to be rendered insoluble must possess chemical groups which can be modified for linkage to the solid support without abolishing or seriously impairing interaction with the complimentary protein. If the strength of interaction of the free complex in solution is very strong, e.g., a K_i of about $10^{-9} M$, a decrease in affinity of 3 orders of magnitude upon preparation of the insoluble derivative may still leave an effective and selective adsorbent. The important parameter is the effective experimental affinity, i.e., that displayed between the protein in solution and the insolubilized ligand under specific experimental conditions. In practice it has been very difficult to prepare insoluble adsorbents with dissociation constants greater than 5 mM under optimal conditions unless the ligand is attached to the carrier backbone by a long arm (Cuatrecasas, 1970b). It is possible in theory to prepare adequate adsorbents for such systems if enough of the inhibitor can be coupled to the solid support, but the porosity of the insoluble support is an important factor as it is possible to have a very large amount of ligand bound to a carrier in such a way that most of it is not freely accessible to the protein.

For successful purification by affinity chromatography the part of the inhibitor critical for interaction with the macromolecule to be purified must be sufficiently distant from the solid matrix to minimize steric interference with binding (Cuatrecasas, 1970b). Steric considerations seem most important in proteins of high molecular weight. The problem may be approached by preparing an inhibitor with a long hydrocarbon chain or arm attached to it. The arm can be attached in turn to the insoluble support (Cuatrecasas *et al.*, 1968). Alternatively, such a hydrocarbon extension arm can be attached to the solid support first (Cuatrecasas, 1970b,c).

The amount of material attached to the solid support must be accurately determined. This is done preferably by determining (by radioactivity, absorbance, amino acid analysis, etc.) the amount of ligand released from the substituted matrix by acid or alkaline hydrolysis. Exhaustive Pronase or carboxypeptidase digestion followed by amino acid analysis has been used in some cases where oligopeptides are attached to carriers. It is also possible to determine the radioactive content of the unhydrolyzed solid support. An alternative means of quantitation is to estimate the amount of ligand which is not recovered in the final washings. But this method is comparatively inaccurate when a very large excess of ligand is added during the coupling procedure, or when appreciable noncovalent adsorption to the solid matrix occurs (demanding large volumes of solvent for thorough washing). It is operationally more useful to express the degree of ligand substitution on the solid matrix in terms of concentration (i.e., as μmoles of ligand per milliliter of packed gel) rather than on the basis of dry weight.

IV. COVALENT LINKING REACTIONS

A. Cellulose and Polystyrene Supports

Perhaps the most common covalent linkage of enzymes and antigens to cellulose and polystyrene derivatives is by the reaction of histidyl and tyrosyl residues of these proteins with diazotized derivatives of the matrix. Pressman et al. (1942) were the first to utilize this technique in binding ovalbumin to sheep erythrocytes using bisdiazobenzidine as a divalent coupling agent. Campbell et al. (1951; Malley and Campbell, 1963) subsequently characterized immunoadsorbents prepared by coupling ovalbumin to diazotized p-aminobenzyl cellulose in detail. Antigens are now commonly coupled to diazonium cellulose (Talmage et al., 1954; Gurvich, 1957; Nezlin, 1959; Gurvich et al., 1959; 1961; Sehon, 1963; Moudgal and Porter, 1963; Gurrich and Drizlikh, 1964; Webb and Lapresle, 1964) and Diazotized poly-p-aminostyrene (Manecke and Gillert, 1955; Gyenes et al., 1958; Kent and Slade, 1960; Gyenes and Sehon, 1960; Yagi et al., 1960; Sehon, 1962; Webb and Lapresle, 1964; Cuatrecasas, 1969a). Antigens have been coupled by their amino groups to carboxymethyl cellulose (amide linkage) with dicyclohexylcarbodiimide in organic solvents (Manecke and Gillert, 1955; Gyemes et al., 1958; Kent and Slade, 1960). Solid supports containing isothiocyanate groups have been used to link proteins by means of their amino groups (Kent and Slade, 1963). Other proteins have been attached through their amino groups to insoluble polystyrene and cellulose supports by reactions involving acyl halides (Isliker, 1953; Jagendorf et al., 1963; Robbins et al., 1967), acid anhydrides (Levin et al., 1964), acyl azides (Micheel and Evers, 1949; Mitz and Summaria, 1961), and sulfonyl chlorides (Isliker, 1953). Nucleotides having monoesterified phosphate groups have been linked to cellulose in organic (Gilham, 1962, 1967) and aqueous (Gilham, 1968) solvents by carbodiimide reagents.

B. Beaded Agarose Supports

Because of the recent interest, and successful use of agarose for affinity chromatography, a detailed description of the various procedures available for coupling ligands and proteins to these supports will be presented. The general schemes will be described now and illustrated with specific examples in later sections. A gentle general method has been developed by Porath et al. for coupling compounds containing primary aliphatic (1970b,c; Axén et al., 1967) or aromatic (Cuatrecasas et al., 1968; Cuatrecasas, 1971b,c; Cuatrecasas and Anfinsen, 1971a) amines to cross-linked polysaccharides such as dextran and agarose (Fig. 1). Agarose is activated by reaction with cyanogen halides at alkaline pH, then washed and coupled with the protein or ligand at acidities varying between pH 6–10 (Cuatrecasas, 1970b; Cuatrecasas and Anfinsen, 1971a). The exact chemical nature of the intermediate formed by cyanogen halide treatment of polysaccharide derivatives is unknown, but the principal products formed upon coupling with amino compounds appear to be derivatives of imino carbonate and isourea (Porath, 1968). Notably, both these groups retain the basicity of the amino group of the ligand after coupling. Therefore, though the charge on the amino group of the ligand may be known to contribute to binding, the agarose-ligand gel may still be effective. Enzymes, antibodies, peptides, and hormone-binding serum proteins have been purified by means of affinity chromatography by attaching ligands or proteins *directly* to agarose using the cyanogen bromide procedures described in detail elsewhere (Cuatrecasas et al., 1968; Cuatrecasas, 1970b,c; Cuatrecasas and Anfinsen, 1971a). Examples will be presented later.

A number of chemical derivatives of agarose have recently been prepared, innovating a variety of methods for attaching ligands and proteins to agarose (Cuatrecasas, 1970b,c). These derivatives, which increase the general versatility of affinity chromatography, should prove especially useful in cases where hydrocarbon arms of varying length are to be interposed between the matrix and the ligand, where amino groups are not present on the ligand, and where it is desirable to remove the intact protein-ligand complex by specific chemical cleavage of the ligand-matrix bond.

Perhaps the most versatile derivative is ω-aminoalkyl-agarose (Fig. 1, C), which can either be utilized directly or can be derivatized. Aliphatic diamines of the general formula $NH_2(CH_2)_x NH_2$ (for example, ethylenediamine, hexamethylenediamine), can be attached by one of the amino groups (with little

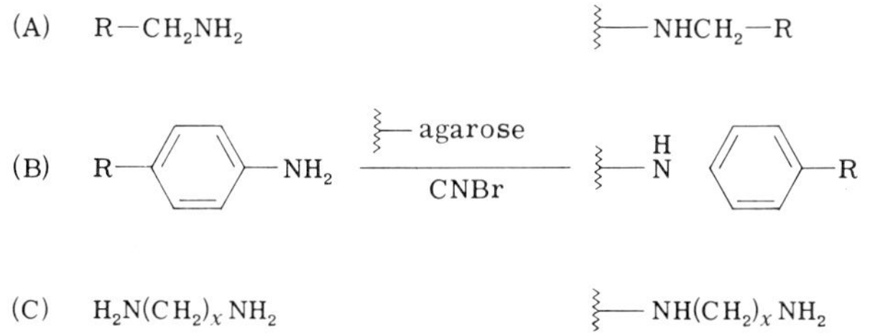

(A) $R-CH_2NH_2$ $-NHCH_2-R$

(B) $R-\langle\bigcirc\rangle-NH_2$ $\xrightarrow[CNBr]{\text{agarose}}$ $-\overset{H}{\underset{}{N}}-\langle\bigcirc\rangle-R$

(C) $H_2N(CH_2)_x NH_2$ $-NH(CH_2)_x NH_2$

Fig. 1. Coupling of ligands containing primary amino groups to agarose by the cyanogen bromide procedure (Cuatrecasas et al., 1968; Jerina and Cuatrecasas, 1970; Cuatrecasas, 1970b,c; Cuatrecasas and Anfinsen, 1971)a.

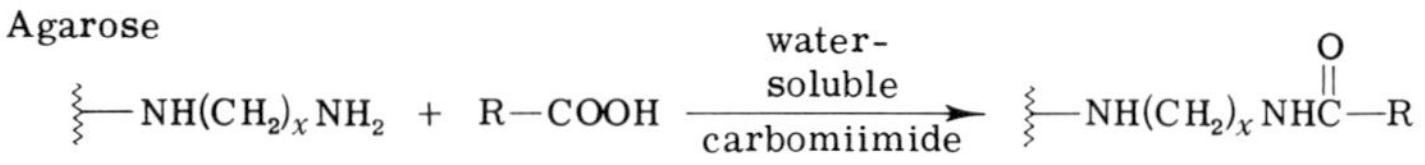

Agarose

$\xi\!-\!NH(CH_2)_x NH_2$ + R—COOH $\xrightarrow[\text{carbomiimide}]{\text{water-soluble}}$ $\xi\!-\!NH(CH_2)_x NH\overset{\overset{O}{\|}}{C}\!-\!R$

Fig. 2. Coupling ligands containing carboxyl groups to ω-aminoalkyl-agarose derivatives (Cuatrecasas, 1970b,c).

cross-linking) by the cyanogen bromide procedure (Cuatrecasas, 1971b,c). In this way it is easy to insert extensions of considerable length, depending on the nature of $(CH_2)_x$. Amino-agarose derivatives are thus obtained for ligand attachment in a variety of ways. All these procedures are performed easily and quickly in aqueous media.

Ligands containing carboxyl groups can be coupled directly to amino-agarose with a water-soluble carbodiimide (Fig. 2) (Cuatrecasas, 1970b,c). Effective adsorbents for sulfhydryl proteins have been prepared by coupling p-chloromercuribenzoate to such derivatives (Cuatrecasas, 1970b). O-Succinyl estradiol has also been coupled to such amino-agarose derivatives (Cuatrecasas, and Puca, 1970). Amino-agarose can be treated in aqueous media, at pH 6, with succinic anhydride to form a long-armed carboxylic-agarose derivative (Fig. 3). Primary aliphatic or aromatic amines can in turn be coupled to these with water-soluble carbodiimides (Fig. 3) (Cuatrecasas, 1970b,c). Adsorbents for staphylococcal nuclease (Cuatrecasas, 1970b,d), bacterial β-galactosidase (Steers et al., 1971), liver cell tyrosine aminotransferase (Miller et al., 1971), and glycerol-3-phosphate dehydrogenase (Holohan et al., 1970) have also been prepared in this way.

Bromoacetyl-agarose is prepared from amino-agarose by reaction in aqueous neutral medium with O-bromoacetyl-N-hydroxysuccinimide (Fig. 4) (Cuatrecasas, 1970b,d). Alkylated agarose derivatives can then be prepared by reaction with proteins, peptides, or ligands containing amino, phenolic, or imidazole substituents. Histidine, estradiol, and antibodies have been attached to agarose by such procedures (Cuatrecasas, 1970b,d).

Diazonium-agarose derivatives can be prepared by treating amino-agarose with p-nitrobenzoyl azide, reducing the nitro group with sodium dithionite, and diazotization with nitrous acid (Fig. 5). Ligands having phenolic or imidazole groups react rapidly with diazonium-agarose. It has been possible by these

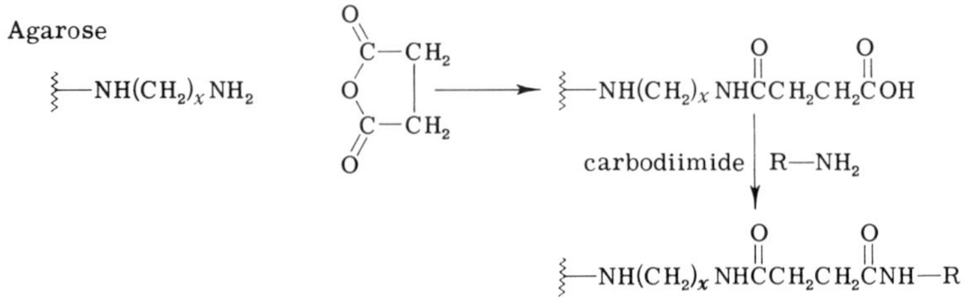

Fig. 3. Preparation and use of carboxyl group-containing agarose by reacting ω-aminoalkyl-agarose with succinic anhydride (Cuatrecasas, 1970b,c).

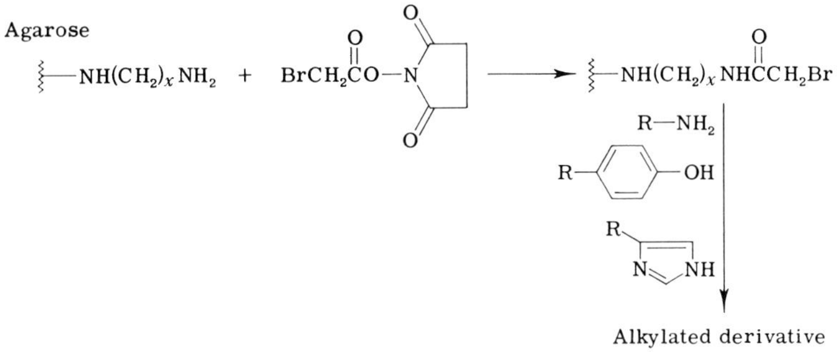

2. Selective Adsorbents Based on Biochemical Specificity 89

$$\text{Agarose} \quad -\text{NH}(\text{CH}_2)_x\text{NH}_2 \;+\; \text{BrCH}_2\overset{O}{\overset{\|}{\text{C}}}\text{O}-\text{N}\begin{pmatrix}O\\\|\\O\end{pmatrix} \longrightarrow -\text{NH}(\text{CH}_2)_x\text{NH}\overset{O}{\overset{\|}{\text{C}}}\text{CH}_2\text{Br}$$

$$\Big\downarrow \begin{array}{l} R-\text{NH}_2 \\ R-\text{C}_6\text{H}_4-\text{OH} \\ R-\text{imidazole} \end{array}$$

Alkylated derivative

Fig. 4. Preparation and use of bromoacetyl-agarose derivatives (Cuatrecasas, 1970b,c).

procedures to attach estradiol, Congo Red dye, histidine, glucagon, and insulin antibodies to agarose (Cuatrecasas, 1970b,d; Cuatrecasas and Puca, 1970).

Sulfhydryl-agarose derivatives can be prepared by treating amino-agarose with N-acetylhomocysteine thiolactone at 4° and pH 9.7 (Fig. 6) (Cuatrecasas, 1970b,c). Ligands containing free carboxyl groups can in turn be coupled to such thiol-agarose derivatives by reaction with water-soluble carbodiimides. The thiol-ester linkage produced can be specifically cleaved by a short exposure

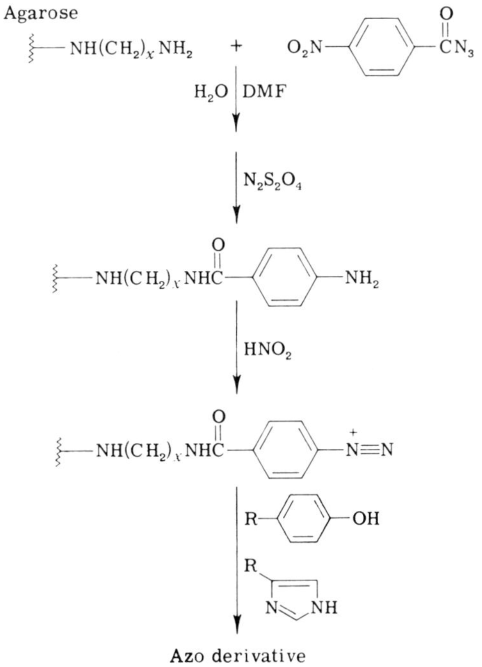

$$\text{Agarose} \quad -\text{NH}(\text{CH}_2)_x\text{NH}_2 \;+\; \text{O}_2\text{N}-\text{C}_6\text{H}_4-\overset{O}{\overset{\|}{\text{C}}}\text{N}_3$$

$$\Big\downarrow \text{H}_2\text{O} \mid \text{DMF}$$

$$\Big\downarrow \text{N}_2\text{S}_2\text{O}_4$$

$$-\text{NH}(\text{CH}_2)_x\text{NH}\overset{O}{\overset{\|}{\text{C}}}-\text{C}_6\text{H}_4-\text{NH}_2$$

$$\Big\downarrow \text{HNO}_2$$

$$-\text{NH}(\text{CH}_2)_x\text{NH}\overset{O}{\overset{\|}{\text{C}}}-\text{C}_6\text{H}_4-\overset{+}{\text{N}}\equiv\text{N}$$

$$\Big\downarrow \begin{array}{l} R-\text{C}_6\text{H}_4-\text{OH} \\ R-\text{imidazole} \end{array}$$

Azo derivative

Fig. 5. Preparation and use of diazonium-agarose derivatives (Cuatrecasas 1970b,c).

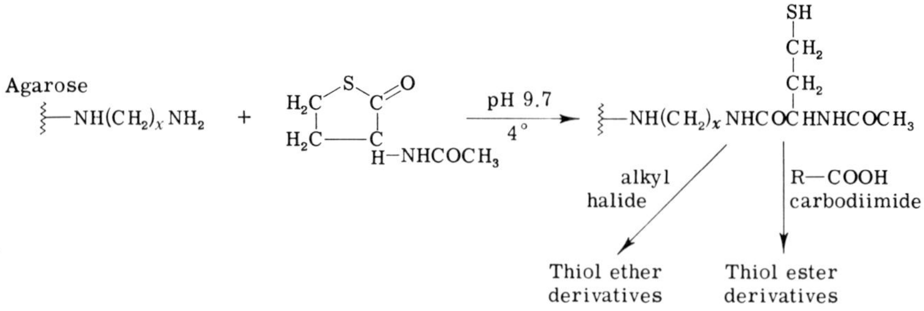

Fig. 6. Preparation af thiol-agarose, and its use in the preparation of ester and ether derivatives (Cuatrecasas, 1970b,c).

to alkaline pH, or with neutral hydroxylamine. O-Succinyl estradiol has been coupled to agarose in this way (Cuatrecasas, 1970b; Cuatrecasas and Puca, 1970).

Ligands that contain aromatic amino groups, which can be diazotized with nitrous acid, can be coupled in azo linkage to a agarose derivative which contains a covalently linked carboxy-terminal tyrosine peptide (Fig. 7). The 3'-p-aminophenyl phosphate derivative of thymidine 5'-phosphate and p-aminophenyl-β-D-thiogalactopyranoside have been attached in this way to agarose to purify staphylococcal nuclease (Cuatrecasas, 1970b) and β-galactosidase (Steers et al., 1971), respectively.

A special advantage of azo-linked ligands is the susceptibility to cleavage of the ligand-agarose bond by reduction with 0.1 M sodium dithionite at

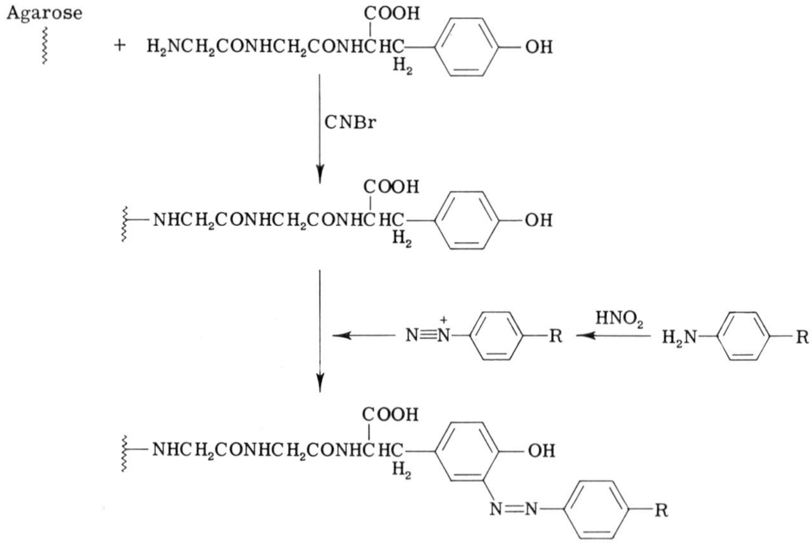

Fig. 7. Preparation of tyrosyl-agarose derivative for coupling diazonium ligands (Cuatrecasas, 1970b,c).

pH 8 (Cuatrecasas, 1970b,c). The ligand-protein complex can thus be removed intact from the solid support which contains the selectively adsorbed protein.

Many of these basic agarose derivatives should be of increasing usefulness in many areas of biochemistry since some of them are now commercially available (Affitron Corp., Rosemead, California).

C. Polyacrylamide Supports

The procedures for polyacrylamide derivatization (Cuatrecasas, 1970b) are based on observations of Inman and Dintzis (1969). These involve conversion of the carboxamide side groups to hydrazide groups (Fig. 8, 1) which in turn

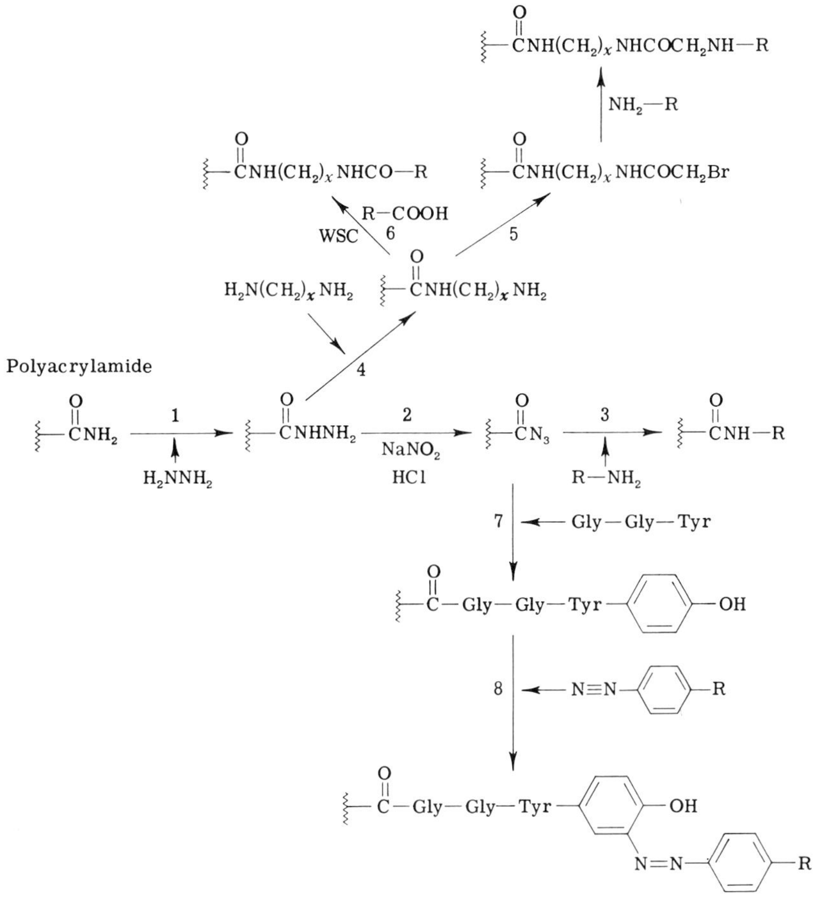

Fig. 8. Various polyacrylamide derivatizations useful in affinity chromatography (Cuatrecasas, 1970b).

are converted with nitrous acid into acyl azide derivatives (Fig. 8, 2). The latter react rapidly with compounds having aliphatic or aromatic primary amino groups without intermediate washings or transfers (Fig. 8, 3).

In general, the specific polyacrylamide adsorbents are more useful when the ligand is attached at some distance from the matrix (Cuatrecasas, 1970b). For this purpose ω-aminoalkyl (Fig. 8, 4), bromoacetamidoethyl (Fig. 8, 5), Gly-

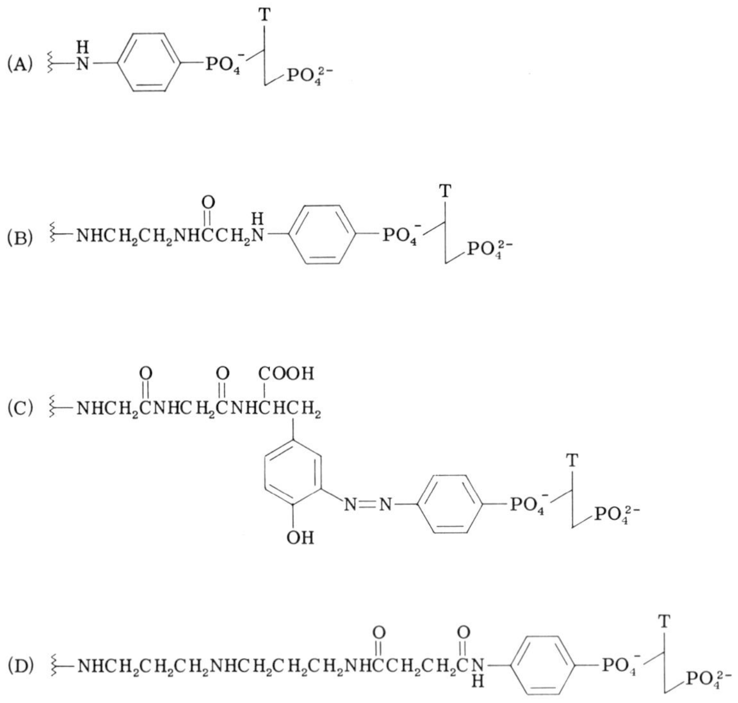

Fig. 9. Specific staphylococcal nuclease affinity chromatographic adsorbents prepared by attaching the competitive inhibitor, pdTp-aminophenol, to various derivatives of agarose or Bio-Gel P-300 (Cuatrecasas, 1970b). In A the inhibitor was attached *directly* to agarose, after activation of the gel with cyanogen bromide, or to polyacrylamide via the acyl azide step. In B ethylenediamine was reacted with cyanogen bromide-activated agarose, or with the acyl azide polyacrylamide derivative. This amino gel derivate was then reacted with N-OH-succinimide ester of bromoacetic acid to form the bromoacetyl derivative; the latter was then treated with the inhibitor. In C the tripeptide, glycylglycyl-tyrosine, was attached by the α-amino group to agarose or polyacrylamide by the cyanogen bromide or acyl azide procedure, respectively; this gel was then reacted with the diazonium derivative of the inhibitor. In D, 3,3′-diaminodipropylamine was attached to the gel matrix by the cyanogen bromide or acyl azide step. The succinyl derivative, obtained after treating the gel with succinic anhydride in aqueous media, was then coupled with the inhibitor with a water soluble carbodiimide. The jagged vertical lines represent the agarose or polyacrylamide backbone. Results of affinity chromatography experiments with these derivatives are summarized in Table I.

Gly-Tyr (Fig. 8, 7), and p-aminobenzamidoethyl derivatives can be prepared by procedures identical to those described for agarose (Cuatrecasas, 1970b). Satisfactory affinity adsorbents have been prepared for the purification of staphylococcal nuclease by many of these procedures (Cuatrecasas, 1971b,d).

As discussed earlier, the principal theoretical advantage of polyacrylamide beads over agarose beads is that a much higher degree of ligand substitution is possible with the former. However, the considerably lower porosity of currently available polyacrylamide beads may limit their use in the affinity chromatography of very large proteins. Furthermore, some shrinkage and decrease in porosity of the gels occurs during formation of the acyl azide intermediate. Thus, it may be possible to prepare an adsorbent containing a very high concentration of ligand, much of which is inaccessible to the protein to be purified. Only the most porous beads (Bio-Gel P-300) have been found effective for the purification of staphylococcal nuclease, a protein with a molecular weight of 17,000 (Cuatrecasas, 1970b). Studies comparing the effectiveness of adsorbents prepared by attaching p-aminophenyl-β-D-thiogalactopyranoside to agarose and to polyacrylamide in the *same* way showed that the polyacrylamide derivatives were very ineffective, despite having 5–10 times more bound inhibitor (Steers et al., 1971). The relative ineffectiveness of polyacrylamide compared to agarose adsorbents is also demonstrated, although less dramatically, with studies of affinity chromatography of staphylococcal nuclease (Fig. 9, Table I).

TABLE I

CAPACITY OF COLUMNS CONTAINING VARIOUS AGAROSE AND POLYACRYLAMIDE ADSORBENTS FOR STAPHYLOCOCCAL NUCLEASE

Derivative[a]	Capacity[b] (mg of nuclease/ml of gel)
Agarose	
A	2
B	8
C	8
D	10
Polyacrylamide	
A	0.6
B	2
C	3

[a] These refer to those depicted in Fig. 9.

[b] Derivatives were diluted with unsubstituted gel to obtain 2 μmoles of ligand/ml of packed gel; the theoretical capacity for staphylococcal nuclease is therefore about 40 mg/ml. (Data from Cuatrecasas, 1970b.)

V. CONDITIONS FOR CHROMATOGRAPHY ON AFFINITY COLUMNS

The buffer and temperature for adsorption of a particular protein are controlled by the specific properties of the protein to be purified. Affinity purification need not be restricted to column procedures. In fact, it may be advisable to use batch purification procedures when small amounts of protein are to be extracted from very crude or particulate protein mixtures using an adsorbent of a very high affinity. Column flow rates in such cases may be slow, and purification might be more expeditiously carried out by adding a slurry of the specific adsorbent to the crude mixture, followed by batchwise washing and elution. In some cases involving complexes of very high affinity, as observed in certain antibody-antigen systems, it may be preferable to adsorb the protein to the solid support and then to wash extensively in a column, eluting the protein by suspending the matrix in an appropriate buffer.

Protein specifically adsorbed to a column of a selective solid carrier will eventually emerge from the column without altering the properties of the buffer if the affinity for the ligand is not too great. With this type of elution, however, the protein is generally obtained in a dilute solution. In most cases proper elution of the protein requires changing the buffer as to pH, ionic strength, or temperature. Dissociation of proteins from adsorbents of very high-affinity may require partially denaturing the protein with guanidine hydrochloride or urea, for example. Ideal elution of a tightly bound protein utilizes a buffer system which causes sufficient alteration in the conformation of the protein to appreciably decrease the affinity of the protein for the ligand, but which is not sufficient to completely unfold the protein. The eluted protein should be neutralized, diluted, or dialyzed immediately to permit prompt reconstitution of the native structure. The effectiveness of such a reconstitution can be tested by rechromatographing the purified protein to determine changes in adsorption to the affinity column. Elution of a specifically adsorbed protein can also be achieved by using a solution containing a high concentration of a specific inhibitor or substrate. The inhibitor can either be the same as the one that is covalently linked to the matrix (and must be used at higher concentrations), or preferably another, stronger, competitive inhibitor. Protein eluted with a buffer containing substrates or inhibitors generally emerges in larger effluent volumes than those obtained when elution is effected by changes in the pH or ionic strength of the buffer.

It is sometimes difficult or impossible to elute proteins which are strongly adsorbed to affinity columns without resorting to the use of extreme conditions of pH or of denaturants (guanidine·HCl, urea) which may irreversibly destroy the biological function of the protein. In such cases it may be of value to remove the intact protein-ligand complex from the solid support. Three of the derivatization products described here are readily adapted to specific chemical cleavage of the ligand–gel bond under relatively mild conditions. The azo-linked derivatives can be cleaved by reduction with sodium dithionite at pH 8.5 (Cuatrecasas, 1970b). This procedure has been of considerable value in studies of the purification of serum estradiol binding protein with derivatives of agarose to which estradiol is attached by the azo linkage (Cuatrecasas, 1970b; Cuatrecasas and Puca, 1970). The serum estradiol binding protein is denatured irreversibly by exposure to pH 3 or 11.5, and by low concentrations of guanidine·HCl (3 M) or urea (4 M). The protein, which binds estradiol very tightly (K_i about 10^{-9} M), can be removed in active form from the agarose-estradiol gel by reductive cleavage of the azo link with dithionite. Carboxylic acid ester derivatives of various ligands have also been linked to agarose by the procedures described here. Such bonds can be cleaved readily by short periods of exposure to pH 11.5 at 4°. Ligands attached to agarose by thiol esters can be similarly cleaved by short exposure to alkaline pH or by treatment with neutral hydroxylamine. These procedures present alternative ways of removing intact ligand-protein complexes from insoluble supports.

VI. IMPORTANCE OF ANCHORING ARMS

The importance of interposing a hydrocarbon chain between the ligand and the agarose backbone was suggested by the relative ineffectiveness of agarose-bound D-tryptophan methyl ester, compared to ε-aminocaproyl-D-tryptophan methyl ester, in the purification of α-chymotrypsin (Cuatrecasas *et al.*, 1968). Several procedures are now available for the preparation of gels which contain hydrocarbon arms of varying length. Figure 9 illustrates several derivatives of agarose and polyacrylamide which contain a competitive inhibitor of staphylococcal nuclease, pdTd-aminophenyl, attached to the solid matrix in various ways (Cuatrecasas, 1970b). Although the affinity adsorbent which contains the ligand attached *directly* to the matrix (Fig. 9,A) is quite effective in selectively extracting the enzyme from dilute aqueous solutions, the protein binding capacity of the gel can clearly be increased by placing the ligand at some distance from the matrix (Table I). However, extending the position of the ligand farther than in the derivative shown in Fig. 9,B does not result in a further increase in binding capacity.

The ligand used in the above examples for the purification of staphylococcal nuclease, pdTp-aminophenyl, is a relatively strong competitive inhibitor (K_i about 10^{-6} M) of this enzyme (Cuatrecasas *et al.*, 1969a). The importance of the arm is much more dramatic in situations which involve ligand-protein interactions of low affinity. This is clearly illustrated in studies of the purification of bacterial β-galactosidase with agarose derivatives containing a relatively weak competitive inhibitor (K_i, about 10^{-3} M), p-aminophenyl-β-D-thiogalactopyranoside (Fig. 10) (Steers et al., 1971). No binding of enzyme was observed

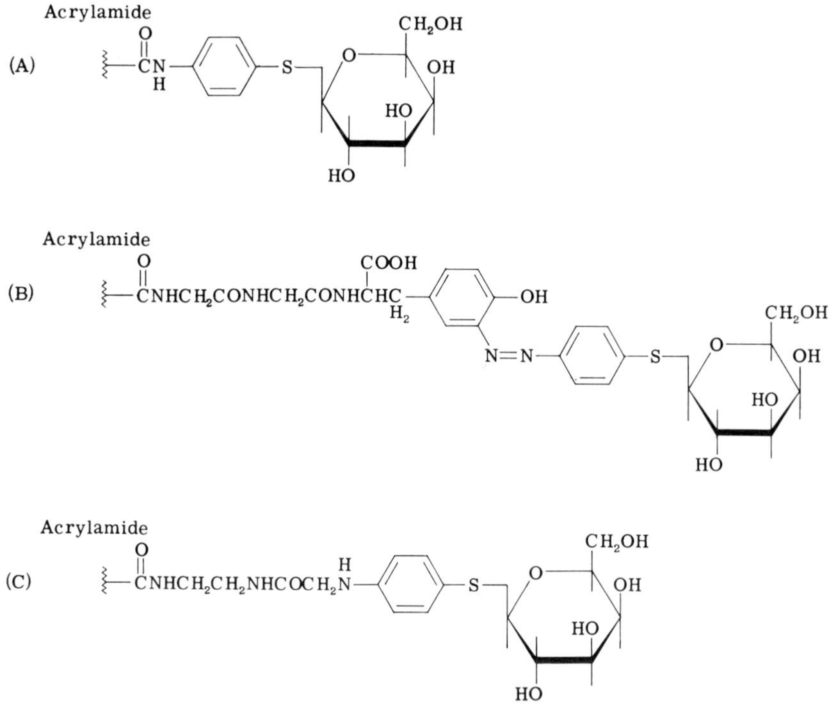

Fig. 10. Affinity chromatographic adsorbents prepared by attaching a weak competitive inhibitor of bacterial β-galactosidase, p-aminophenyl-β-D-thiogalactopyranoside, to agarose through extensions of varying length (Pollard et al., 1971). The ligand was attached directly to cyanogen bromide activated agarose (A), to the bromoacetaminoethyl derivative (B), and to the succinylated 3,3′-diaminodipropylamine derivative (C), respectively. Columns containing these gels exhibit no β-galactosidase binding (derivative A), only a slight retardation in the chromatographic migration of the enzyme (derivative B), and very strong binding of the protein (derivative C) (Steers et al., 1971).

with derivatives of agarose (or polyacrylamide) which contained the ligand attached directly to the matrix (Fig. 10,A), even though derivatives were prepared which contained as much as 14 μmoles of inhibitor per milliliter of packed gel. By placing the inhibitor at a moderate distance (about 10 Å) from the solid support (Fig. 10,B) the β-galactosidase activity emerges slightly behind the major protein breakthrough in column chromatography experiments. However, agarose derivatives obtained after insertion of a very long arm (about 20 Å) between the agarose matrix and ligand (Fig. 10,C) adsorb β-galactosidase very strongly from various sources (Steers et al., 1971).

2. Selective Adsorbents Based on Biochemical Specificity

Similarly, ineffective adsorbents for muscle glycerol-3-phosphate dehydrogenase result if glycerol-3-P is attached directly to the agarose backbone (Holohan et al., 1970). Attachment of 1-Cl- and 1-Br- analogs of the same ligand to hexamethylenediamine-treated agarose results in effective adsorbents.

Tyrosine aminotransferase from mouse hepatoma tissue culture cells does not bind to agarose derivatives having pyridoxamine phosphate attached directly to the matrix backbone (Fig. 11, upper) (Miller et al., 1971). In contrast,

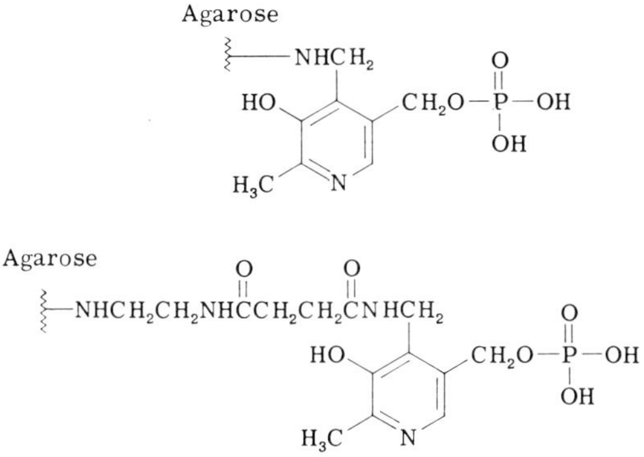

Fig. 11. Specific absorbents for purification of hepatoma tissue culture cell tyrosine aminotransferase (Miller et al., 1971). When pyridoxamine phosphate is attached directly to the agarose backbone (upper), the enzyme does not bind at all to affinity columns containing this adsorbent. Attachment of the same ligand by a long arm (lower) results in a very effective adsorbent for this enzyme.

very strong and specific adsorption of this enzyme occurs when the ligand is attached by a long arm (Fig. 11, lower).

The interposition of such hydrocarbon arms between gel and ligand appears to be most important for interacting systems of low affinity, and in cases involving multisubunit or high molecular weight proteins. The dramatic effects of increasing distance observed in some of these cases may be in part explained by relief of steric restrictions imposed by the matrix backbone, and perhaps by the increased flexibility and mobility of the ligand as it protrudes farther into the solvent.

VII. SPECIFIC ADSORBENTS FOR PROTEIN PURIFICATION

Table II describes a variety of enzymes and other proteins which have been purified by affinity chromatography. Antigen and antibody purifications have been omitted from this table. A large number of enzymes have now been purified by affinity chromatography with the use of inhibitors, substrates, effectors and

TABLE II

Proteins That Have Been Purified by Affinity Chromatography[a]

Enzyme or protein	Ligand	Insoluble matrix	Reference
Tyrosinase	Aminophenol	Cellulose	Lerman (1963)
Flavokinase	Flavins	Cellulose	Arsenis and McCormick (1964, 1966)
Avidin	Biotinyl chloride	Cellulose	McCormick (1965)
Avidin	Biocytin	Agarose	Cuatrecasas and Wilchek (1968)
Staphylococcal nuclease	pdTp-Aminophenyl	Agarose	Cuatrecasas et al. (1968); Cuatrecasas (1970b)
Staphylococcal nuclease	pdTp-Aminophenyl	Polyacrylamide	Cuatrecasas (1970b)
Active site modified nuclease	pdTp-Aminophenyl	Agarose	Cuatrecasas et al. (1969b); Cuatrecasas (1970a)
α-Chymotrypsin	ε-Aminocaproyl-D-tryptophan methyl ester	Agarose	Cuatrecasas et al. (1968)
	4-Phenylbutylamine	Agarose	Stevenson and Landman (1971)
Carboxypeptidase A	L-Tyrosine-D-tryptophan	Agarose	Cuatrecasas et al. (1968)
β-Galactosidase	p-Aminophenylthiogalactoside	Agarose, polyacrylamide	Steers et al. (1971)
β-Galactosidase	p-Aminophenylthiogalactoside	Cross-linked bovine gamma globulin	Tomino and Paigen (1970)
Tyrosine aminotransferase	Pyridoxamine phsophate	Agarose	Miller et al. (1971)
Flavin-linked glycerol-3-phosphate dehydrogenase	1-Cl- and 1-Br-G-3-P	Agarose	Holohan et al. (1970)
Pancreatic ribonuclease A	p-Aminophenyl-p-U-cP	Agarose	Wilchek and Gorecki (1969)

2. Selective Adsorbents Based on Biochemical Specificity

Pyridoxal kinase	Adenosine diphosphate	Agarose	Neary and Diven (1970)
Acetylcholinesterase	Substrate analogs	Agarose	Berman and Young (1971)
Wheat proteases	Hemoglobin	Agarose	Chua and Bushuk (1969)
Sulfhydryl proteins (hemoglobin, thyroglobulin)	p-Chloromercuribenzoate	Agarose	Cuatrecasas (1970b); Salvatore and Cuatrecasas (1969)
Sulfhydryl proteins	3,6-Bis(acetatomercurimethyl)-dioxane	Cross-linked dextran	Eldjarn and Jellum (1963)
Mercaptopapain	p-Aminophenylmercuric acetate	Agarose	Sluyterman and Widjenes (1970)
Chorismate mutase	L-Tryptophan	Agarose	Sprossler and Lingens (1970)
Papain	Gly-Gly-(O-benzyl)-L-Tyr-L-Arg	Agarose	Blumberg et al. (1969)
3-Deoxy-D-arabinoheptulosonate-7-phosphate synthetase	L-Tyrosine	Agarose	Chan and Takahashi (1969)
Estradiol-binding serum protein	Various estradiols	Agarose	Cuatrecasas (1970d)
Human amyloid protein	Congo Red dye	Agarose, polyacrylamide	Cuatrecasas (1970d)
Synthetic ribonuclease-S-peptide	S-Protein	Agarose	Kato and Anfinsen (1969)
Synthetic analogs of P₂ (staphylococcal nuclease)	P₃ Peptide	Agarose	Ontjes and Anfinsen (1969)
Thyroxine-binding serum protein	L-Thyroxine	Agarose	Pensky and Marshall (1969)
Plasminogen	L-Lysine	Agarose	Deutsch and Mertz (1970)
Affinity labeled protein peptides	Enzyme or antibody	Agarose	Givol et al. (1970)
Rat liver ribonuclease inhibitor	Ribonuclease	Carboxymethyl cellulose	Gribnau et al. (1970)
Isoleucyl-tRNA	Specific synthetase	Agarose	Denburg and DeLucca (1970)

[a] Cases of antibody or antigen purification have been omitted from this table.

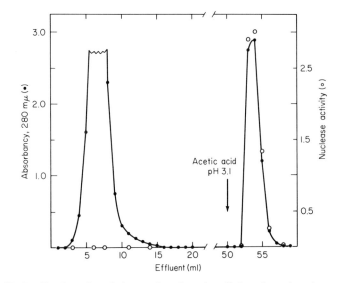

Fig. 12. Purification of staphylococcal nuclease by affinity adsorption chromatography on a nuclease-specific agarose column. The column was equilibrated with 0.05 M borate buffer, pH 8.0, containing 0.01 M $CaCl_2$. Approximately 40 mg of partially purified material containing about 8 mg of nuclease was applied in 3.2 ml of the same buffer. After 50 ml of buffer had passed through the column, 0.1 M acetic acid was added to elute the enzyme. Nuclease, 8.2 mg, and all of the original activity was recovered. The flow rate was about 70 ml per hour. (Data from Cuatrecasas et al., 1968.)

cofactors (Table II). Figure 12 shows a typical column chromatographic pattern; in this case staphylococcal nuclease was purified with the adsorbent shown in Fig. 9,A.

Studies of the functional effects of chemical modifications of purified enzymes frequently reveal incomplete loss of enzymic activity. It is often difficult to determine if this activity represents residual native enzyme or an altered protein with diminished catalytic power. Separation of the active native and the catalytically inert protein is often difficult, but in certain cases the difficulty may be resolved by using affinity adsorbents. For example, modification of staphylococcal nuclease by attachment of a single molecule of an affinity labeling reagent through an azo linkage to an active site tyrosyl residue resulted in loss of 83% of the enzyme activity (Cuatrecasas, 1970a). Chromatography of this enzyme solution on a nuclease-specific agarose column revealed that the residual activity was due entirely to a 20% contamination of native enzyme; complete resolution of the two components was possible by affinity chromatography (Fig. 13) (Cuatrecasas, 1970a). Similar separations were made with partially active preparations of staphylococcal nuclease modified with bromoacetamidophenyl affinity labeling reagents (Cuatrecasas et al., 1969b), and residual native

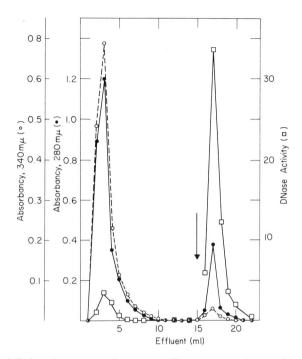

Fig. 13. Affinity chromatography on specific agarose column of staphylococcal nuclease treated with a 1.7-fold molar excess a diazonium labeling reagent derived from pdTp-aminophenyl. About 3 mg of chemically modified nuclease, containing 17% of the DNase activity of the native enzyme, were applied to a 0.5 × 7 cm column which contained agarose conjugated with the inhibitor, pdTp-aminophenyl (0.8 μmoles per ml of agarose). The column was equilibrated and developed with 0.05 M borate buffer, pH 8.0, and 10 mM CaCl$_2$. The bound enzyme was eluted with NH$_4$OH, pH 11 (arrow). The small amount of enzymic activity present in the early, unretarded peak could be removed by rechromatography of this peak through the same column. The specific activity of the small amount of protein adsorbing strongly to the column was identical with that of native nuclease. (Data from Cuatrecasas, 1970a.)

nuclease could be separated from an inactive peptide fragment obtained by specific tryptic cleavage (Ontjes and Anfinsen, 1969).

A number of noncatalytic proteins, such as estradiol (Cuatrecasas, 1970d) and thyroxine (Pensky and Marshall, 1969; Pages et al., 1969) serum-binding proteins, avidin (McCormick, 1965; Cuatrecasas and Wilchek, 1968), and amyloid protein (Cuatrecasas, 1970d), have been purified by affinity chromatography by using cyanogen bromide procedures (Cuatrecasas et al., 1968).

Affinity chromatography has also been used to purify peptides prepared by organic synthesis. A polypeptide corresponding to the amino acid sequence from residue 6 through 47 in staphylococcal nuclease (149 amino acids), which is obtained by organic synthesis, can be purified by passing the crude synthetic polypeptide mixture through a column containing an agarose-linked native peptide which comprises residues 49–149 (Taniuchi and Anfinsen, 1969). In solution these two peptides separately have no discernable tertiary structure or enzymic activity, but upon mixing there is regeneration of enzymic activity. Advantage of the affinity of these two peptides is taken to obtain a "functional" purification of the synthetic peptide. Similarly, it is possible to purify synthetic RNase-S-peptide derivatives on agarose columns containing RNase-S-protein (Kato and Anfinsen, 1969). The specific activity of the purified peptide is increased by this procedure, but the chemical and catalytic properties of the purified materials suggest the presence of closely related synthetic side products which bind tightly to the S-protein conjugate but which yield enzymically inactive complexes with RNase-S-protein.

No attempt will be made here to extensively review the field of immunoadsorbents. In the last few years the use of agarose for such derivatives has received much attention, and it is now the preferred carrier for this purpose in many laboratories, for the same reasons that make this carrier nearly ideal for enzyme purification. Agarose immunoadsorbents [(cyanogen bromide procedures, described in Cuatrecasas *et al.* (1968); Cuatrecasas, 1970b)] have been used successfully to purify insulin antibodies (Cuatrecasas, 1969b), a mouse myeloma IgA protein which binds nitrophenyl ligands (Goetzl and Metzger, 1970), a variety of antihapten antibodies (Wofsy and Burr, 1969), staphylococcal nuclease antibodies (Omenn *et al.*, 1970), canine brucellosis antibodies (Howe *et al.*, 1970), and antipolysaccharide antibodies from streptococcal Group A antiserum (Parker and Briles, 1970).

Antibodies have also been attached to agarose to purify various antigens. Insulin antibodies attached to agarose are very effective in purifying small quantities of insulin (Fig. 14) (Cuatrecasas, 1970d; Akanuma *et al.*, 1970). agarose-coupled antibody to human chorionic somato-mammotropin has been used to purify labeled and unlabeled hormone for radioimmunoassay (Weintraub, 1970).

Insoluble derivatives of proteins can also be used advantageously to study effects of subunit aggregation and polymerization. For example, insulin, a polypeptide hormone with a great propensity for aggregation, is completely active biologically when coupled to agarose (Cuatrecasas, 1969a). Since these agarose derivatives are washed extensively with urea and guanidine before use, it is clear that it is the monomeric species of insulin which is the biologically active form of the hormone. This point must be carefully considered in deducing structure-function relations of this protein from the tertiary structure obtained by X-ray crystallography, since the crystallographic data are obtained from crystals of the hexamer.

Another example of the use of an insoluble protein is given by a study of agarose-coupled subunits of glutamine synthetase (Hennig and Ginsburg, 1970). The binding of ions, substrates, or inhibitors to the dissociated subunits cannot be studied because the subunits cannot be maintained in the dissociated state in the absence of denaturants. The agarose-coupled subunits are catalytically inactive. If the sulfhydryl groups of these subunits are blocked with p-chloromercuriphenylsulfonate they can not be adenylylated by ATP : glutamine synthetase adenylyltransferase, while adenylylation proceeds quite normally if the organomercurial is removed with mercaptoethanol.

In addition to attaching proteins or peptides directly through their amino groups to agarose by the cyanogen bromide procedure, it is possible to use the bromoacetyl, diazonium, or sulfhydryl derivatives of agarose (Cuatrecasas

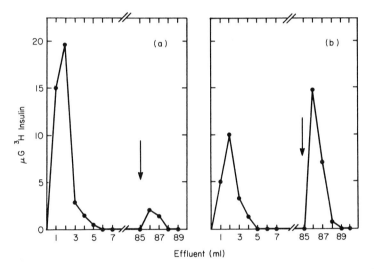

Fig. 14. Affinity chromatography patterns of ^{3}H-insulin on columns containing purified antiporcine insulin sheep immunoglobulin which had been coupled by the cyanogen bromide procedure to agarose at pH 9 (a) and at pH 6.5 (b). Forty mg of purified γ-globulin, in 10 ml of 0.2 M sodium citrate, pH 6.5, or 8 ml of 0.2 M sodium bicarbonate, pH 9.0, were added to 5 ml of agarose activated with 300 mg of cyanogen bromide per milliliter of gel. Based on the recovery of absorbancy (280 mμ) in the washes, about 90–95% of the protein was coupled in both cases. The specific adsorbent was washed with 20 times its volume of 6 M guanidine · HCl before use. ^{3}H-Acetyl insulin, 40 μg, in 1 ml of 0.05 M borate buffer at pH 8.5 containing 0.1 N NaCl, and 0.5% bovine albumin, was applied to a 12 × 0.6 cm column. The column was then washed with 85 ml of the above buffer, and elution of bound insulin was accomplished with 6 M guanidine · HCl (arrow). The theoretical capacity for insulin binding, based on the amount of antibody covalently linked and on the capacity of the antibody to bind insulin in solution, was about 30 μg for both adsorbents. Column A adsorbed 2 μg of insulin, or 7% of its theoretical capacity, whereas column B adsorbed 23 μg, or 77% its theoretical capacity. (Data from Cuatrecasas, 1970d).

1970b) described above. Proteins attached in the latter fashion will extend some distance from the matrix backbone by an arm, which may be very useful in overcoming steric difficulties when interactions with other macromolecules are being studied. For example, the study of the interaction of certain insolubilized proteins or polypeptides with intact cells or cell structures may best be achieved with such derivatives.

An important consideration in the covalent attachment of a biologically active protein to an insoluble support is that the protein should be attached to the matrix by the fewest number of bonds possible (Cuatrecasas, 1969a, 1970a). This will increase the probability that the attached macromolecule will retain its native tertiary structure, and its properties may more nearly resemble those of the native protein in solution. Proteins react with cyanogen bromide-activated agarose through the unprotonated form of their free amino groups. Since most proteins are richly endowed with lysyl residues, most of which are exposed to solvent, it is likely that such molecules will have multiple points of attachment to the resin when the coupling reaction is done at pH 9.5 or higher, as is usually the case. This problem may be circumvented by carrying out the coupling procedure at a less favorable pH. For example, it has been demonstrated that if antibodies are coupled to agarose at pH 6.0–6.5 the resultant immunoadsorbent has a much greater capacity for antigen than that which is prepared by performing the coupling procedure at pH 9.5 (Cuatrecasas, 1970b,d). The studies on anti-insulin shown in Fig. 14 illustrate the importance of attaching large proteins by few linkage points. The immunoglobulin coupled to activated agarose at pH 9 could adsorb only 7% of its theoretical capacity for insulin, whereas that coupled at pH 6.5 could bind 77% of its theoretical capacity. Since the total protein content of both derivatives is the same, the former derivative must contain much immunoglobulin which is incapable of binding antigen effectively.

VIII. SPECIFIC NUCLEIC ACID ADSORBENTS

Water-insoluble derivatives of nucleic acids have been used in a number of interesting studies. It has been shown that single stranded DNA binds to nitrocellulose filters (Aliapoulios et al., 1965), and that this DNA can bind homologous RNA (Goldhaber, 1965). This property has been exploited to develop general procedures for isolation of gene-specific mRNA (Nyggard and Hall, 1963; Bautz and Reilly, 1966).

Oligonucleotides have been covalently attached by this hydroxyl group to acetylated phosphocellulose (phosphodiester bond) with dicyclohexylcarbodiimide in pyridine (Adler and Rich, 1962) or methanol (Bautz and Holt, 1962). Coupling of oligonucleotides by a terminal phosphate to the hydroxyl group of cellulose has been done very successfully with special carbodiimide reagents both in anhydrous (Gilham, 1962; Gilham and Robinson, 1964; Cozzarelli *et al.*, 1967; Jovin and Kornberg, 1968) and aqueous (Gilham, 1968) media. These insoluble oligo- and polynucleotide celluloses have been used to separate, fractionate, and determine the structure of various nucleic acids (Gilham and Robinson, 1964; Erhan *et al.*, 1965; Sander *et al.*, 1966). These studies are based on the relative stabilities of complexes formed between components in the aqueous mixture and complementary insoluble polynucleotide structures.

Water-insoluble polynucleotide celluloses have also been useful in the study of enzyme mechanisms. Cozzarelli *et al.* (1967) attached a polynucleotide to cellulose by a phosphate terminus. Enzymes that catalyze the covalent linking of interrupted deoxyribonucleotide strands of a bihelix could then be assayed by determining the amount of ^{3}H-poly dC (initially free in solution) which became linked covalently to the polynucleotide cellulose. Jovin and Kornberg (1968) studied various polynucleotide cellulose derivatives as solid state primers and templates for DNA polymerases.

Blasi *et al.* (1969) have attached purified histidyl-tRNA-synthetase (lacking other acyl tRNA-synthetase activities) by the cyanogen-bromide procedure (pH 6.5 for coupling step) described by Cuatrecasas for protein coupling (1970b). Columns containing this adsorbent strongly adsorb histidyl-tRNA which can then be eluted with 1 M NaCl.

Very recently techniques have been described by Poonian *et al.* (1971) for coupling nucleic acids to agarose by using the cyanogen bromide procedure described by Cuatrecasas (1970b). They demonstrated that single-stranded deoxy- and ribonucleic acid sare readily linked, whereas double-stranded ones are not. They described procedures by which single-stranded ends could be introduced into double-stranded molecules, thus permitting their covalent attachment to agarose. DNA polymerase from HeLa cells adsorbed to DNA agarose about fifty times more efficiently that to DNA cellulose.

IX. ISOLATION OF CELLS, RECEPTOR STRUCTURES, AND OTHER PARTICULATE CELL STRUCTURES

Insolubilized ligands or proteins can also be used to study interactions of various molecules with intact cells and to explore membrane phenomena. Although the macromolecules involved were not attached *covalently* to the supporting matrix, the recent studies of Wigzell and Andersson (1969) on the fractionation of immunologically active cells on columns of glass or plastic beads to which antigen coatings were tightly adsorbed indicate the power of the approach. Similar studies on the immunoadsorption of cells on reticulated polyester polyurethane foam coated with antibodies have been reported by Evans *et al.*

Mage and Peterson (1969). Very recently Davie and Paul (1970) have utilized various haptene-agarose derivatives to fractionate immunocompetent lymphoid cells from guinea pigs. Truffa-Bachi and Wofsy (1970) have recently found that affinity columns prepared by attaching haptenes to large polyacrylamide beads can selectively bind cells which produce antihaptene antibodies of corresponding specificity. These results indicate that affinity columns may be useful in purifying cell populations.

Insulin-agarose derivatives are capable of stimulating several different biological properties of intact isolated fat cells nearly as effectively as does native insulin (Fig. 15) (Cuatrecasas, 1969a). This indicates that the principal inter-

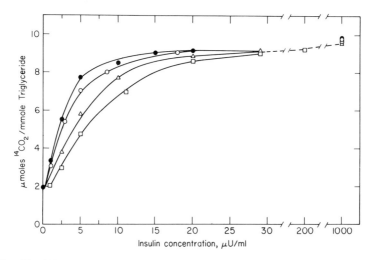

Fig. 15. Effect of insulin-agarose concentration on the oxidation of glucose-U-^{14}C by isolated adipose cells. Fat cells (3 to 8 μmoles of triglyceride) were incubated for 2 hours at 37° in 2 ml of medium containing 0.5 mM glucose-U^{14}C (0.1 μCi/μmole). The data presented are for native insulin (●), insulin coupled to agarose (at pH 9) through lysine B29 (○), insulin coupled (at pH 5) through phenylalanine B1 (△), and insulin acetylated at positions A1 and B1 are coupled to agarose by lysine B29 (□). (Data from Cuatrecasas, 1969a.)

action of insulin is with cell surface structures and that the structural requirements for this interaction are essentially intact in the insulin-agarose derivative. Columns prepared with these insulin-agarose derivatives, in contrast to those with unsubstituted agarose, bind intact fat cell "ghosts" very strongly (Cuatrecasas, 1970d). The latter can be specifically eluted intact from the insulin-agarose columns with concentrated insulin solutions. These studies indicate that such insoluble hormone derivatives interact very effectively with membrane structures, and thus are promising tools for the isolation of "receptor" proteins for hormones, drugs, etc. Furthermore, it may be possible to separate discrete cell populations according to a specific function, the capacity to interact specifically with a given ligand or protein.

Pyridoxamine phosphate agarose derivatives (Fig. 11, lower), which bind tyrosine aminotransferase from hepatoma tissue culture cells very strongly, have been used to partially purify ribosomes which are capable after elution of specifically synthesizing immunologically and enzymically identifiable tyrosine aminotransferase (Miller et al., 1971). The partially synthesized, nascent protein present on such ribosomes may recognize the insoluble ligand. This may be a procedure which may be of general use in ribosome purification, particularly in cases of multisubunit enzymes.

References

Adler, A. J., and Rich, A. (1962). *J. Amer. Chem. Soc.* **84**, 3977.
Akanuma, Y., Kuquya, T., and Hayashi, M. (1970). *Biochem. Biophys. Res. Commun.* **38**, 947.
Aliapoulios, M. A., Savery, A., and Munson, P. L. (1965). *Fed. Prod., Fed. Amer .Soc. Exp. Biol.* **24**, 322.
Arsenis, C., and McCormick, D. B. (1964). *J. Biol. Chem.* **239**, 3093.
Arsenis, C., and McCormick, D. B. (1966). *J. Biol. Chem.* **241**, 330.
Axén, R., Porath, J., and Ernbäck, S. (1967). *Nature (London)* **214**, 1302.
Bautz, E. K. F., and Holt, B. D. (1962). *Proc. Nat. Acad. Sci. U.S.* **48**, 400.
Bautz, E. K. F., and Reilly, E. (1966). *Science* **151**, 328.
Berman, J. D., and Young, M. (1971). *Proc. Nat. Acad. Sci. U.S.* **68**, 395.
Blasi, F., Goldbreger, R. F., and Cuatrecasas, P. (1969). Unpublished data.
Blumberg, S., Schechter, I., and Berger, A. (1969). *Isr. J. Chem.* **7**, 125.
Campbell, D. H., Leuscher, E., and Lerman, L. S. (1951). *Proc. Nat. Acad. Sci. U.S.* **37**, 575.
Chan, W. C., and Takahashi, M. (1969). *Biochem. Biophys. Res. Commun.* **37**, 272.
Chua, G. K., and Bushuk, W. (1969). *Biochem. Biophys. Res. Commun.* **37**, 545.
Cozzarelli, N. R., Melechen, M. E., Jovin, T. M., and Kornberg, A. (1967). *Biochem. Biophys. Res. Commun.* **28**, 578.
Cuatrecasas, P. (1969a). *Proc. Nat. Acad. Sci. U.S.* **63**, 450.
Cuatrecasas, P. (1969b). *Biochem. Biophys. Res. Commun.* **35**, 531.
Cuatrecasas, P. (1970a). *J. Biol. Chem.* **245**, 574.
Cuatrecasas, P. (1970b). *J. Biol. Chem.* **245**, 3059.
Cuatrecasas, P. (1970c). *Nature (London)* **228**, 1327.
Cuatrecasas, P. (1970d). Unpublished data.
Cuatrecasas, P., and Anfinsen, C. B. (1971a). *Methods Enzymol.* **21** (in press).
Cuatrecasas, P., and Anfinsen, C. B. (1971b). *Annu. Rev. Biochem.* **40**, in press.
Cuatrecasas, P., and Illiano, G. (1971). *J. Biol. Chem.* (in press).
Cuatrecasas, P., and Puca, G. A. (1970). Unpublished data.
Cuatrecasas, P., and Wilchek, M. (1968). *Biochem. Biophys. Res. Commun.* **33**, 235.
Cuatrecasas, P., Wilchek, M., and Anfinsen, C. B. (1968). *Proc. Nat. Acad. Sci. U.S.* **61**, 636.

Cuatrecasas, P., Wilchek, M., and Anfinsen, C. B. (1969a). *Biochemistry* **8**, 2277.
Cuatrecasas, P., Wilchek, M., and Anfinsen, C. B. (1969b). *J. Biol. Chem.* **244**, 4316.
Davie, J. M., and Paul, W. E. (1970). *Cell. Immunol.* **1**, 404.
Denburg, J., and De Lucca, M. (1970). *Proc. Nat. Acad. Sci. U.S.* **67**, 1057.
Deutsh, D. G., and Mertz, E. T. (1970). *Science* **170**, 1095.
Eldjarn, L., and Jellum, E. (1963). *Acta Chem. Scand.* **17**, 2610.
Erham, S. L., Northrup, L. G., and Leach, F. R. (1965). *Proc. Nat. Acad. Sci. U.S.* **53**, 646.
Evans, W. H., Wage, M. G., and Peterson, E. A. (1969). *J. Immunol.* **102**, 899.
Franklin, E. C. (1964). *Progr. Allergy* **8**, 58.
Gilham, P. T. (1962). *J. Amer. Chem. Soc.* **84**, 1311.
Gilham, P. T. (1964). *J. Amer. Chem. Soc.* **86**, 4982.
Gilham, P. T. (1968). *Biochemistry* **7**, 2809.
Gilham, P. T., and Robinson, W. E. (1964). *J. Amer. Chem. Soc.* **86**, 4985.
Gilham, P. T., and Robinson, W. E. (1964). *J. Amer. Chem. Soc.* **86**, 4985.
Givol, D. Weinstein, Y., Gorecki, M., and Wilchek, M. (1970). *Biochem. Biophys. Res. Commun.* **38**, 825.
Goetzl, E. J., and Metzger, H. (1970). *Biochemistry* **9**, 1267.
Goldhaber, P. (1965). *Science* **147**, 407.
Gribnau, A. A. M., Schoenmakers, J. G. G., van Kraaikamp, M., and Bloemendol, H. (1970). *Biochem. Biophys. Res. Commun.* **38**, 1064.
Gurvich, A. E. (1957). *Biochemistry (USSR)* **22**, 977.
Gurvich, A. E., and Drizlikh, G. I. (1964). *Nature (London)* **203**, 648.
Gurvich, A. E., Kapner, R. B., and Nezlin, R. S. (1959). *Biochemistry (USSR)* **24**, 129.
Gurvich, A. E., Kuzovlena, O B., and Tumanova, A. E. (1961). *Biochemistry (USSR)* **26**, 803.
Gyenes, L., and Sehon, A. H. (1960). *Can. J. Biochem. Physiol.* **38**, 1235 and 1249.
Gyenes, L., Rose, B., and Sehon, A. H. (1958). *Nature (London)* **181**, 1465.
Hennig, S. B., and Ginsburg, A. (1970). Personal communication.
Hjerten, S. (1962). *Arch. Biochem. Biophys.* **99**, 446.
Holohan, P. D., Mahajan. K., and Fondy, T. P. (1970). *Fed. Proc., Fed. Amer. Soc. Exp. Biol.* **29**, 888 (abstr.); personal communication (1970).
Howe, C. W., Morisset, R., and Spink, W. W. (1970). *Fed. Proc., Fed. Amer. Soc. Exp. Biol.* **29**, 830 (abstr.).
Inman, J. K., and Dintzis, H. M. (1969). *Biochemistry*, **8**, 4074.
Isliker, H. C. (1953). *Ann. N.Y. Acad. Sci.* **57**, 225.
Jagendorf, A. T., Patchornik, A., and Sela, M. (1963). *Biochim. Biophys. Acta.* **78**, 516.
Jerina, D. M., and Cuatrecasas, P. (1970). *Proc. Int. Congr. Pharmacol., 4th, Basel Switzerland, 1969*. Vol. I, p. 236. Schwabe, Basel/Stuttgart.
Jovin, T. M., and Kornberg, A. (1968). *J. Biol. Chem.* **243**, 250.
Katchalski, E. (1962). *Polyamino Acids, Polypeptides, Proteins, Proc. Int. Symp., 1st, 1961* p. 283.
Kato, I., and Anfinsen, C. B. (1969). *J. Biol. Chem.* **244**, 1004.
Kent, L. H., and Slade, J. H. R. (1960). *Biochem. J.* **77**, 12.
Kent, L. H., and Slade, J. H. R. (1963). *Proc. Int. Congr. Biochem., 5th, 1961* Vol. 9, p. 474.
Lerman, L.S. (1963). *Proc. Nat. Acad. Sci. U.S.* **39**, 232.
Levin, Y., Pecht, M., Goldstein, L., and Katchalski, E. (1964). *Biochemistry* **3**, 1905.
McCormick, D. B. (1965). *Anal. Biochem.* **13**, 194.
Malley, A., and Campbell, D. H. (1963). *J. Amer. Chem. Soc.* **85**, 487.

Manecke, G. (1962). *Pure Appl. Chem.* **4**, 507.
Manecke, G., and Gillert, K. E. (1955). *Naturwissenschaften* **42**, 212.
Micheel, F., and Evers, J. (1949). *Makromol. Chem.* **3**, 200.
Miller, J. D., Cuatrecasas, P. and Thompson, E. B. (1971). *Proc. Nat. Acad. Sci. U.S.*, **68**, 1014.
Mitz, M. A., and Summaria, L. J. (1961). *Nature (London)* **189**, 576.
Moudgal, N. R., and Porter, R. R. (1963). *Biochim. Biophys. Acta* **71**, 185.
Neary, J. T., and Diven, W. F. (1970). *J. Biol. Chem.* **245**, 5585.
Nezlin, R. S. (1959). *Biochemistry (USSR)* **24**, 282.
Nyggard, A. P., and Hall, B. D. (1963). *Biochem. Bophys. Res. Commun.* **12**, 98.
Omenn, G. S., Ontjes, D., and Anfinsen, C. B. (1970). *Nature (London)*, **225**, 189.
Ontjes, D. A., and Anfinsen, C. B. (1969). *J. Biol. Chem.* **244**, 6316.
Pages, R. A., Cahnmann, H. J., and Robbins, J. (1969). Personal communication.
Parker, D. C., and Briles, D. (1970). *Fed. Proc., Fed. Amer. Soc. Exp. Biol.* **29**, 438 (abstr.).
Pensky, J., and Marshall, J. S. (1969). *Arch. Biochem. Biophys.* **135**, 304.
Poonian, M. S., Schlabach, A. J., and Weissbach, A. (1971). *Biochemistry*, **10**, 424.
Porath, J. (1968). *Nature (London)* **218**, 834.
Porath, J., Axén, R., and Ernbäck, S. (1967). *Nature (London)* **215**, 1491.
Porter, R. R., and Press, E. M. (1962). *Annu. Rev. Biochem.* **31**, 625.
Pressman, D., Campbell, D. H., and Pauling, L. (1942). *J. Immunol.* **44**, 101.
Robbins, J. B., Haimovich, J., and Sela, M. (1967). *Immunochemistry* **4**, 11.
Salvatore, G., and Cuatrecasas, P. (1969). Unpublished data.
Sander, E. G., McCormick, D. B., and Wright, L. D. (1966). *J. Chromatogr.* **21**, 419.
Sehon, A. H. (1962). *Pure Appl. Chem.* **4**, 483.
Sehon, A. H. (1963). *Brit. Med. Bull.* **19**, 183.
Silman, I. H., and Katchalski, E. (1966). *Annu. Rev. Biochem.* **35**, 873.
Sluyterman, L. A. E., and Widjenes, J. (1970). *Biochim. Biophys. Acta* **200**, 593.
Sprossler, B., and Lingens, F. (1970). *FEBS Lett.* **6**, 232.
Steers, E., Cuatrecasas, P., and Pollard, H. (1971). *J. Biol. Chem.* **246**, 1960.
Stevenson, K. J., and Landman, A. (1971). *Canadian J. Biochem.* **49**, 119.
Talmage, D. W., Baker, H. R., and Akeson, W. (1954). *J. Infec. Dis.* **84**, 199.
Taniuchi, H., and Anfinsen, C. B. (1969). *J. Biol. Chem.* **244**, 3864.
Tomino, A., and Paigen, K., (1970). *In* "The Lac Operon" (D. Zipser and J. Beckwith eds.), p. 233. Cold Spring Harbor Laboratory, Cold Spring Harbor, New York.
Truffa-Bachi, P., and Wofsy, L., *Proc. Nat. Acad. Sci. U.S.* (1970). *Proc. Nat. Acad. Sci. U.S.* **66**, 685.
Webb, T., and Lapresle, C. (1964). *Biochem. J.* **91**, 24.
Weetall, H. H., and Hersh, L. S. (1969). *Biochim. Biophys. Acta.* **185**, 464.
Weintraub, B. (1970). *Biochem. Biophys. Res. Commun.* **39**, 83.
Wide, L., Axén, R. and Proath, J. (1967). *Immunochemistry* **4**, 381.
Wigzell, H., and Andersson, B. (1969). *J. Exp. Med.* **129**, 23.
Wilchek, M., and Gorecki, M. (1969). *Eur. J. Biochem.* **11**. 491.
Wofsy, L., and Burr, B. (1969). *J. Immunol.* **103**, 380.
Yagi, Y., Engel, K., and Pressman, D. (1960). *J. Immunol.* **85**, 375.

CHAPTER 3   Solid Phase Synthesis: The Use of Solid Supports and Insoluble Reagents in Peptide Synthesis

GARLAND R. MARSHALL AND R. B. MERRIFIELD

I.	Introduction	111
	A. Solid Phase Synthesis	112
	B. Strategy	114
II.	Polymeric Carboxy Protection	115
	A. Modified Benzyl Esters	118
	B. Phenyl Esters	126
	C. Alkyl Esters	131
	D. Benzhydryl Polymeric Support	136
	E. Amide Support	137
III.	Polymeric Amino Protection	138
IV.	Polymeric Side-Chain Protection	141
V.	Polymeric Coupling Reagents	143
	A. Polymeric Active Esters	144
	B. Polymeric Carbodiimide	148
VI.	Effects of Physical Properties of Polymeric Supports	148
VII.	Automation	155
VIII.	Similar Developments in Conventional Synthesis	156
IX.	Difficulties in Analysis	158
X.	Results	160
XI.	Conclusions and Summary	162
	References	162

To promise the System is a serious thing.
Søren Kierkegaard, *Philosophical Fragments*

I. INTRODUCTION

The ability to duplicate synthetically the compounds which are produced in living organisms has been a goal of chemistry since Wöhler first accomplished this feat with urea. Only in the last few decades has the intricate complexity of some of nature's polymers come to light. With this knowledge has come the desire to produce these compounds synthetically and to subsequently modify their structures in order to study the relationship between structure and activity.

Methods for determining the unique structure of amino acid polymers, peptides and proteins, were developed early (Sanger, 1946), and for this reason the subsequent development of synthetic procedures has progressed farther than that for nucleic acids or polysaccharides. Based on the classical work of Fischer, Curtius, Abderhalden, and Bergmann, duVigneaud (duVigneaud et al., 1953) synthesized for the first time a peptide hormone, oxytocin, a work for which he was subsequently awarded the Nobel Prize. Further developments of protecting groups, coupling reagents, and measurements of racemization led to a culmination of the classical approach—the synthesis of a 39-residue chain, ACTH, by Schwyzer and Sieber in 1963 and the two chains of insulin, 21 and 30 residues, (Meienhofer et al., 1963; Katsoyannis et al., 1964; Kung et al., 1965) in 1965. These represented major accomplishments involving tremendous efforts by many individuals. These chains seem relatively small when considering the length of proteins which have now been sequenced, e.g. chymotrypsin, carboxypeptidase.

In order to approach problems of this magnitude as well as to make the synthesis of analogs of smaller structures easier, the techniques of classical synthesis had to be streamlined and simplified. One of the most time-consuming steps is the purification of intermediates since this usually requires a unique procedure for each step. A method which would unify the purification procedures so that each intermediate was purified in the same way would be a vast improvement over normal methods. Such a procedure has been developed which not only unifies the purification scheme, but reduces it to simple washing operations. This procedure, solid phase peptide synthesis, has been the subject of several reviews (Merrifield, 1965a,b, 1967, 1968a, 1969a; Ferriere, 1966; Halstrom, 1967; Okuda, 1968; Vesa, 1968) and, more recently, a book (J. M. Stewart and Young, 1969).

A. Solid Phase Synthesis

The basic idea of solid phase synthesis, which is the attachment of one of the reactants of a bimolecular reaction to a solid support with subsequent purification by filtration, was conceived by Merrifield in 1959. It is generally applicable in principle throughout synthetic chemistry and is especially useful when synthesizing polymers of known sequence from different monomer units of a single type, i.e., specific-sequence heteropolymers. Several variations of the original application have been attempted and others which might offer significant advantages will be pointed out in the course of the review. Solid phase synthesis, in its present state, offers many practical advantages in the synthesis of peptides.

The potential of this approach, however, is only beginning to be realized, and the various aspects which have been explored only suggest part of its possible scope.

The original procedure as outlined in Fig. 1 requires the attachment of one end of the desired polymer sequence to the solid support through a covalent link.

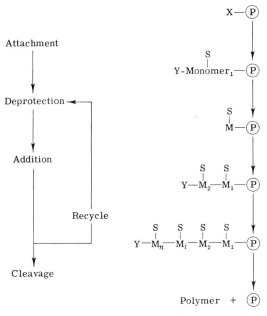

Fig. 1. General outline of solid phase synthesis. M, monomer; P, polymeric support; Y, protecting group removed during synthesis; and S, side-chain protecting group.

This link must be sufficiently stable to allow all subsequent steps to be completed without its cleavage, and yet be sufficiently labile to be split at the end of synthesis under conditions which will not destroy the integrity of the chosen polymer. Obviously, the site on the monomer unit which forms the covalent link must be either activated or exposed while other reactive sites are blocked. This allows the desired link to be formed specifically.

One of the blocked groups must then be exposed selectively so that a second monomer unit can be coupled. The other potentially reactive sites on the monomer units must remain completely blocked, and the subsequent coupling of a new monomer unit must be as nearly quantitative as possible. Intermediate purification is achieved by washing the polymer support free of reagents and by-products, and the success of this simplified approach is dependent on driving the deblocking and coupling reactions very near to completion. The exposure of reactive sites and the coupling of additional units is repeated in a stepwise manner until the desired sequence is established. The oligomer is then removed from the polymer support by cleaving the covalent link between them.

The potential advantages to such an approach are numerous. It eliminates the losses due to physical manipulations since all the reactions can be carried out in a single vessel. The purification of intermediates is a routine washing procedure which eliminates the losses, both in time and material, that normally result during the development of appropriate purification schemes for each particular intermediate. The requisite use of excess reagents to force the individual reactions to completion favors a high yield of final product.

Disadvantages are also numerous depending upon the implementation. The polymer support imposes difficulties in assaying the extent of reaction. Tactics for selection of protecting groups are not independent. The choice of the link to the support limits the choice of the removable blocking group on each monomer unit. In turn, this limits the protecting groups that are to remain throughout the synthesis. Unless the reactions near completion, numerous abortive sequences will build up and the purification of such a mixture of one-residue deletions may prove exceedingly difficult. Several ways to overcome this problem have been devised. One way to minimize the problem involves a strategy similar to that of the classical fragment methodology in which the units used for the stepwise synthesis are actually oligomers. This increases the differences between the desired chain and those chains missing a single unit and allows easier purification. It thus appears possible in practice to minimize the disadvantages of the solid phase approach in order to reap its benefits.

B. Strategy

To borrow a term from Bodanszky and Ondetti (1966), peptide chemistry can be dichotomized into strategy, which is the general synthetic approach, and tactics, which are the particular chemical means by which the strategy is implemented. While solid phase synthesis may be considered a general strategy in its own right, the possible approaches are sufficiently great to warrant a discussion of the substrategies of solid phase synthesis.

Basically, the strategies can be broken down according to which component of the peptide chain is linked to the polymer support. The initial procedure which has been the most highly developed is to link the amino acid through the carboxyl group of the C-terminal residue. The major advantage to this scheme is that stepwise elongation of the chain is possible through activation of protected amino acid residues, a procedure known to minimize the possibility of racemization.

Alternatively, the chain can be linked to the support through the amino group of the amino-terminal residue, as was originally demonstrated by Letsinger and

Kornet (1963). The major disadvantage of this approach is the severe limitation imposed on possible coupling reagents because of the dangers of racemization during activation of the growing peptide chain. A third possibility could be to attach the polymer through an appropriate side-chain functional group such as an —SH of cysteine. One would then have the possibility of lengthening the chain in either direction depending on the position of the cysteinyl residue in the desired peptide. Another approach has been the use of polymeric activating agents as exemplified by a polymeric active ester (Fridkin *et al.*, 1965a; Wieland and Birr, 1966), which can be charged with the carboxyl component and the desired peptide released when the polymeric activated carboxyl component is then reacted with the amino component.

For the purposes of this review, the various developments in the use of solid supports and reagents in peptide synthesis will be classified according to the "strategy" of solid phase, i.e., which function the polymeric material is serving: as amino, carboxyl, or side-chain protection and/or as activating reagent. The tactics that have been used will be indicated under the appropriate strategy. In addition, various common features of the several approaches will be discussed separately. These include polymeric properties, automation, analyses on polymers, and homogeneity of reactive sites.

It should be stated that one of the initial goals of the solid phase method was to devise a set of tactics that would be compatible with all of the amino acid residues in a wide range of sequences. This goal is not easily achieved, however; due, in part, to the rather severe limitations imposed on the other protecting groups by the choice of the group to be removed after each addition of monomer. Although special purpose supports and tactics are becoming more prevalent, the criteria of compatibility will be used in the evaluation of the various approaches so far investigated.

II. POLYMERIC CARBOXYL PROTECTION

The choice of attachment to the support through the carboxyl group of the C-terminal amino acid reflects the prevailing philosophy of petide chemistry. The choice of strategy in classical peptide chemistry is either to build up small segments and combine them or to add amino acids stepwise at the amino end of the chain. While the latter course is more desirable from the point of view of racemization, the former is more practical for purification as it yields a product which remains quite different from the starting materials. Since the solid phase method is built on the premise of reacting under conditions in which the growing chain reacts quantitatively (sometimes not fully realized in practice), the purification problems normally associated with stepwise synthesis do not hinder the application of this strategy to solid phase synthesis. The alternative approach of coupling two large, slowly diffuable reactants, one of which is a polymeric bead, would not appear too desirable a task. When given a choice between extending a chain by coupling a protected amino acid as the carboxyl component or amino component, the choice is obvious due to the decreased sensitivity to racemization of N-protected amino acids versus N-protected peptides.

The original, and most highly developed, procedure is that of Merrifield (1962) in which the covalent link to the polymer support was a benzyl ester (actually a modified *p*-alkyl benzyl ester). The most commonly used polymer, styrene cross-linked with divinylbenzene, was chloromethylated with chloromethyl ether under Friedel-Crafts conditions. The reactions are shown in Fig. 2. While it is possible to substitute all the aromatic groups of the polymer, the substitution is usually limited to 1 to 2 mmoles/g. This was then reacted with the N-protected amino acid to give the ester link. The usual degree of substitution for successful syntheses has been in the range of 0.2 to 0.6 mmoles amino acid/g. Deprotection with 1 N HCl/HOAc, followed by neutralization with triethylamine, and coupling with a Boc amino acid was repeated until the desired sequence was established. The peptide was removed from the support by treatment with HBr in trifluoroacetic acid. Although this procedure has proved very successful (Merrifield, 1969a), numerous modifications and improvements have been suggested and implemented as problems have arisen. It was calculated, and later shown by autoradiography (Merrifield and Littau, 1968), that the growing peptide chains are distributed throughout the gel matrix in the polystyrene-DVB beads. It has been necessary to swell the gel matrix by an appropriately chosen solvent in order to facilitate diffusion throughout the polymer bead. An alternative would be to increase the surface area so that diffusion into the matrix was not so critical. Preliminary investigation (Merrifield, 1965a) into macroporous (Millar *et al.*, 1963) and macroreticular (Kunin *et al.*, 1962) resins which are rigid copolymers containing large pores and much larger surface areas showed a decreasing extent of reaction as the peptide chain was lengthened. The use of pellicular resins (Horvath *et al.*, 1967) and brush resin (Halasz and Sebastian, 1969) in solid phase synthesis has been suggested by Bayer (1969) and Bayer *et al.*, (1970a, b).

Limitation of the growing peptide chains to the more accessible sites of the beads might be possible by limiting chloromethylation or by esterification with a sterically hindered acid followed by replacement with the amino acid as demonstrated by Tilak (1968). Heterogeneity in reactive sites has been indicated by kinetic measurement with the normally prepared supports. Further studies are necessary to evaluate improvements which might be expected from a more homogeneous population of reactive sites.

In one case, difficulty in coupling was overcome by changing to 1% DVB cross-linked polymer from the normal 2% cross-linked support. The logical extreme of this modification is that of Shemyakin *et al.* (1965) who used linear

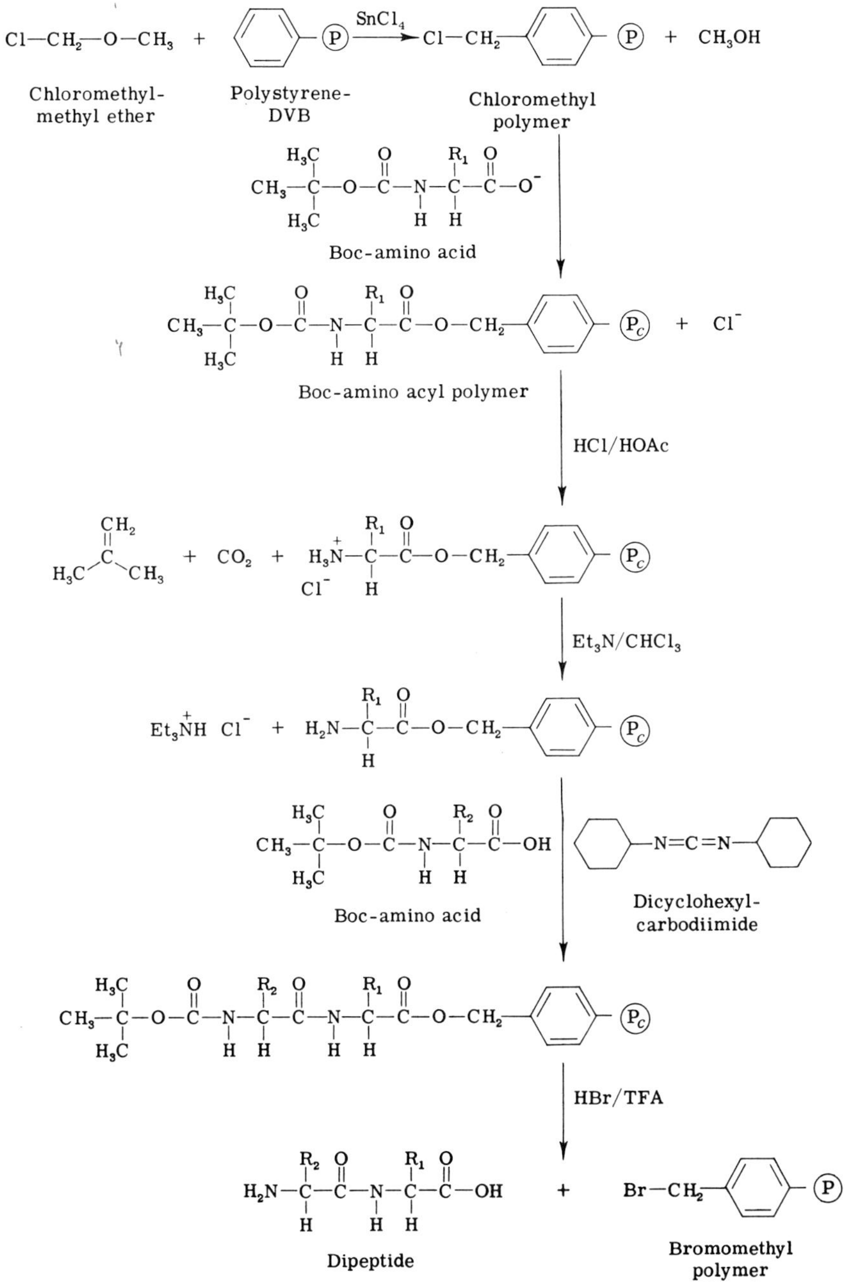

Fig. 2. Outline of the most prevalent procedure for solid phase peptide synthesis. P_c, polymeric support used for carboxyl protection.

polystyrene as a suppport, which could be dissolved for coupling reactions and precipitated for purifications. A difficulty with this approach was the problem of precipitating the support in a form which was easily filtered since the nature of the support changed as the reactions proceeded. Other difficulties noted with this procedure were cross-linking by excess chloromethyl groups and the need for several precipitations to insure purification (Garson, 1968).

Because the majority of work has been done with polymeric carboxyl protection, the effects of the physical properties of the polymer on reactions have been more apparent with this approach. Further discussion of these effects is presented in Section V. It is anticipated that further investigations into this aspect of solid phase synthesis as well as the development of more acid-labile polymeric supports will contribute greatly to the rational application of solid phase synthesis to problems of biological importance.

A. Modified Benzyl Esters

The covalent link to the support which has been most thoroughly explored is a modified benzyl ester in which the aromatic rings of the polystyrene are involved. This p-alkyl benzyl ester has been formed by esterification of the triethylamine salt of the protected amino acid with the chloromethylated polymer [Eq. (1)]. The reaction was carried out in ethanol (Merrifield, 1964b), benzene,

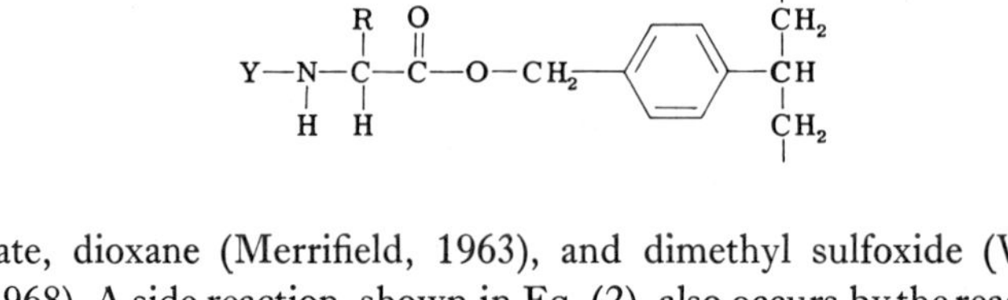

$$
\begin{array}{c}
\text{R O} \\
\text{Y—N—C—C—O}^{-}\ \text{Et}_3\text{NH}^{+} \quad + \quad \text{Cl—CH}_2\text{—}\langle\text{—}\rangle\text{—CH} \\
\text{H H}
\end{array}
\quad \xrightarrow{} \quad
\text{Y—N—C—C—O—CH}_2\text{—}\langle\text{—}\rangle\text{—CH} \tag{1}
$$

ethyl acetate, dioxane (Merrifield, 1963), and dimethyl sulfoxide (Wildi and Johnson, 1968). A side reaction, shown in Eq. (2), also occurs by the reaction with

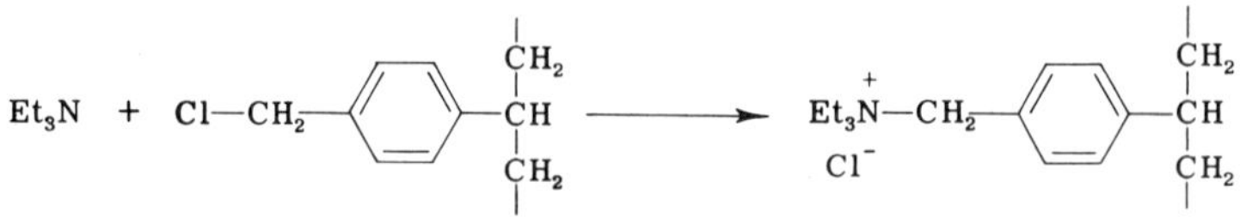

$$\text{Et}_3\text{N} + \text{Cl—CH}_2\text{—}\langle\text{—}\rangle\text{—CH} \longrightarrow \text{Et}_3\overset{+}{\text{N}}\text{—CH}_2\text{—}\langle\text{—}\rangle\text{—CH} \quad \text{Cl}^{-} \tag{2}$$

the tertiary amine, namely quaternization. The amount of quaternary amine can be measured either by total nitrogen analysis (Merrified, 1969a; Gut and Rudinger, 1968; Beyerman et al., 1967) or Volhard titration of chloride displaced from the polymer by excess nitric acid (Merrifield, 1969a). The remaining chloride, which is assumed to be present in unreacted chloromethyl groups, could be determined by Parr fusion or by displacement from the polymer by boiling in pyridine, followed by chloride titration. The quaternary by-product should not interfere with the remaining synthetic steps since it stays as a neutral salt. The use of a more sterically hindered base such as ethyl diisopropylamine, which Weygand (1968) used in the esterification with the phenacyl support, has also been shown to minimize quaternary formation (Merrifield, 1969a). Another side reaction of the chloromethyl groups involves alkylation of Boc-im-benzyl histidine (J. M. Stewart and Young, 1969) and Boc-methionine (Sieber and Iselin, 1968) during the esterification reaction. By prolonged treatment with triethylammonium acetate most of the chloromethyl groups could be acetylated (Merrifield, 1963; Kessler and Iselin, 1966; Dijkstra et al., 1967; Beyerman et al., 1967), but this is normally omitted because interference by unreacted chloromethyl groups with later steps in the synthesis has not been shown, although this possibility should be considered.

Similar conditions of esterification have been used with bromomethyl and iodomethyl polystyrene (Merrifield, 1969a; Tilak, 1968). These derivatives arose from the cleavage of esters formed by the above procedures with either HBr or HI in an attempt to select sterically favored positions on the support. It was shown that the bromomethyl derivative of the polymer, derived from the polymer esterified with a sterically hindered acid, could be quantitatively reesterified while that derived from acetoxymethyl resin only reacted partially with a sterically hindered acid (Tilak, 1968).

To avoid the above possible difficulties of direct esterification, hydroxymethyl polymers have been synthesized, usually by the method of Bodanszky and Sheehan (1966) where the acetoxymethyl derivative is formed from the chloromethyl resin by reaction with potassium acetate in benzyl alcohol. Saponification in alcoholic sodium hydroxide gives the hydroxymethyl resin [Eq. (3)] which can then be coupled with amino acids by use of an appropriate activating

$$CH_3-\overset{O}{\underset{\|}{C}}-O-CH_2-\!\!\left\langle\!\!\bigcirc\!\!\right\rangle\!\!-\!\!\boxed{P}$$

$$\Big\downarrow NaOH/ROH$$

$$CH_3-\overset{O}{\underset{\|}{C}}-OH \quad + \quad HO-CH_2-\!\!\left\langle\!\!\bigcirc\!\!\right\rangle\!\!-\!\!\boxed{P}$$

(3)

agent such as *N,N'*-carbonyl-diimidazole (Bodanszky and Sheehan, 1966), DCCI (Bodanszky and Sheehan, 1966; Inukai *et al.*, 1968) or by a novel method using *N,N*-dimethylformamide-dineopentylacetal (Schreiber, 1967). Similar resins have been used by others (Schreiber, 1967), but the unreacted alcoholic groups precluded their use with DCCI coupling. Acetylation of the free hydroxyls with acetic anhydride (J. M. Stewart and Young, 1969) removed this restriction.

Another approach, shown in Fig. 3 by Dorman and Love (1969), has been the use of a better leaving group on the benzylic carbon of the support. Chloromethyl resin was reacted with dimethyl sulfide to give the dialkyl sulfonium

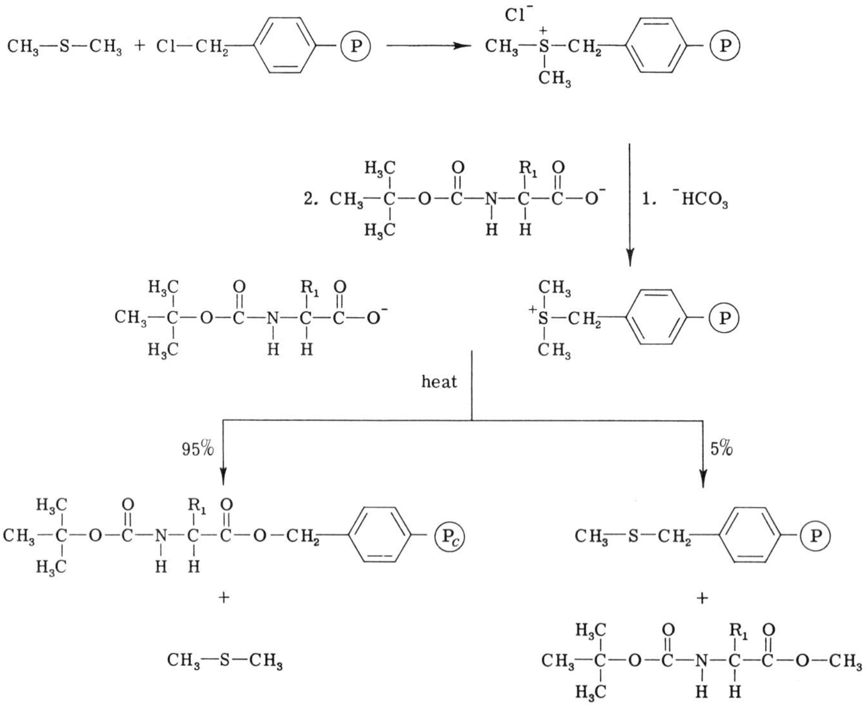

Fig. 3. Esterification of the initial amino acid residue through displacement of the dialkyl sulfonium group according to Dorman and Love (1969).

group. Exchange of the chloride with bicarbonate gives the basic sulfonium ion which will bind a carboxylic acid irreversibly. Upon heating at 80°–85° for 4–5 hours the ester is formed in 87–96% yield. A side reaction gives small amounts of alkyl esters which can be removed by washing and a corresponding amount of alkyl sulfide on the resin which presumably is not detrimental. This approach would offer the advantages that only a stoichiometric amount of N-protected amino acid is needed, that it proceeds without racemization, and that resin esters of high capacity can be easily prepared if desired.

The benzyl ester bond to the support has been shown to be compatible with the use of Boc protecting groups with exposure of the amino function by 1 N HCl/HOAc (Merrifield, 1964a), 4 N HCl/dioxane (J. M. Stewart and Woolley, 1966), TFA (Kusch, 1966; Anfinsen et al., 1967; M. Manning, 1968; Takashima et al., 1968), 50% TFA/CH_2Cl_2 (Gutte and Merrifield, 1969), and 20% TFA/CH_2Cl_2 (Krumdieck and Baugh, 1969). Removal of the Boc group by heat or by formic acid (Halpern and Nitecki, 1967) has not been reported with solid phase synthesis. The synthesis of long peptide chains such as ribonuclease (Gutte and Merrifield, 1969) has pointed out that slight lability of the covalent link accumulates in a rather large loss at the end of a multistep synthesis. The nitrophenylsulfenyl group which can be used with removal either by very dilute acid (Najjar and Merrifield, 1966) or by nucleophilic displacement (Kessler and Iselin, 1966) offers the advantage of much milder removal conditions than the Boc group, thus eliminating cleavage of the peptide chain from the resin support. The enamine, 1-benzoylisopropenyl (Bip), group of Southard et al. (1969) and the biphenylisopropyloxycarbonyl group (Bpoc) of Sieber and Iselin (1968) would also offer such an advantage.

Selection of the side-chain protecting groups is limited to those compatible with the selected amino-protecting group and its method of removal. The side-chain protecting groups which are used must be stable to the conditions which are applied repeatedly in order to expose the amino function for further coupling. Removal of the completed chain from the support has been effected by many of those reagents known to modify benzyl esters. These are outlined in Table I. Until the application of the HF cleavage procedure of Sakakibara et al. (1965, 1967a) to solid phase synthesis by Lenard and Robinson (1967), the standard method used hydrogen bromide according to Ben-Ishai and Berger (1952), but substituted trifluoroacetic acid for the acetic acid solvent to prevent acetylation of serine and threonine hydroxyls (Guttman and Boissonnas, 1959). It should be noted that trifluoroacetic acid alone gave 11% cleavage of angiotensinyl-bradykinin in 1 hour at 25°. At 75° cleavage was complete in 1 hour. Due to the enhancement of the susceptibility of the resin esters to acidic cleavage by the para substituent (F. H. C. Stewart, 1967) small losses of peptide also occur during the acidic deprotection steps. HCl in an appropriate solvent might also be useful for cleavage from the support at elevated temperatures. Anhydrous, liquid hydrogen fluoride has become the method of choice (Lenard and Robinson, 1967) for cleavage from a benzyl ester support. This is due to the fact that an HF-removable group is known for each of the trifunctional amino acids (Sakakibara et al. 1967a,b, 1968b), so that the side-chain protecting groups can be removed simultaneously with cleavage from the resin support. In addition, the method would appear to offer decidedly fewer side reactions when working with

TABLE I

Cleavage Methods for Peptides Bound to Support through Benzyl Ester Linkage

Reagent	Product	Reference
I. Acidolysis		
HBr-HOAc	RCO_2H	Merrifield (1962, 1963); Kusch (1966)
HBr-TFA	RCO_2H	Merrifield (1964a, b; Stewart and Woolley (1965a); Marshall and Merrifield (1965); Marglin and Merrifield (1966)
TFA	RCO_2H	Merrifield (1967)
HF	RCO_2H	Lenard and Robinson (1967); Robinson and Kamen (1968); Gutte and Merrifield (1969)
II. Saponification		
NaOH	RCO_2^-	Merrifield (1963); Shchukina and Skylarov (1966)
NaOEt	RCO_2^-	Loffet (1967)
$NaHCO_3$	RCO_2^-	Harrison and Harrison (1967)
III. Aminolysis		
NH_3	$RCONH_2$	Bodanszky and Sheehan (1964); Beyerman et al. (1967); Manning (1968); Bayer and Hagenmaier (1968)
NH_2NH_2	$RCONHNH_2$	Bodanszky and Sheehan (1964); Kessler and Iselin (1966); Stryer and Haugland (1967); Ohno and Anfinsen (1967); Beyerman and Maassen van den Brink-Zimmermanova (1968)
NH_2OH	$RCONH_2OH$	Beyerman and Maassen van den Brink-Zimmermanova (1968)
NH_2CH_3	$RCONHCH_3$	Beyerman and Maassen van den Brink-Zimmermanova (1968); Takashima et al. (1970)
$NH(CH_3)_2$	$RCON(CH_3)_2$	Takashima et al. (1970)
IV. Transesterification		
Et_3N-R'OH	RCOR'	Lombardo and Piasio (1968); Beyerman et al. (1968)
NH_3-R'OH	RCOR'	Bodanszky and Sheehan (1964); Beyerman et al. (1968)
AER-R'OH	RCOR'	Halpern et al. (1968); Pereira et al. (1969)

tryptophanyl peptides (G. R. Marshall, 1968). An inexpensive apparatus for carrying out the reaction in a closed system has recently been described (Pourchot and Johnson, 1969), similar to the original design of Sakakibara et al. (1967a).

Saponification with 0.2 N NaOH in aqueous ethanol was used by Shchukina and Skylarov (1966) to remove Boc-Gly-Ala-Phe-Val-Gly from the polymer. His-Phe-Arg-Trp-Gly was removed with sodium ethoxide in ethanol by Loffet (1967). Ammonolysis which gives the C-terminal amide [Eq. (4)] was first used

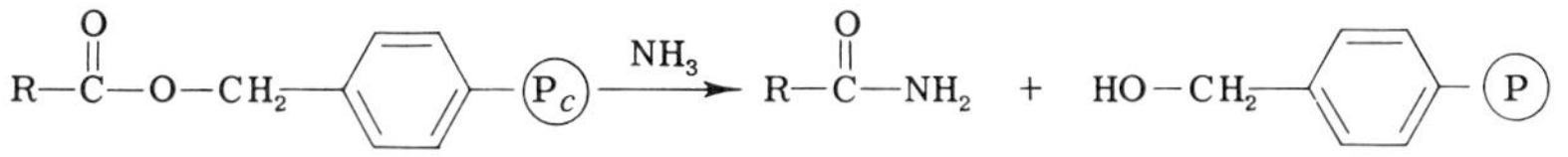

$$R-\overset{O}{\underset{}{C}}-O-CH_2-\langle\rangle-\textcircled{P}_C \xrightarrow{NH_3} R-\overset{O}{\underset{}{C}}-NH_2 + HO-CH_2-\langle\rangle-\textcircled{P} \quad (4)$$

by Bodanszky and Sheehan (1966) who pointed out that side-chain esters will also react with ammonia and should not be present during the cleavage. Although they later experienced difficulty and actually isolated a pure methyl ester of a peptide due to transesterification of the benzyl ester in a saturated solution of NH_3 in methanol, this was attributed to the steric hindrance of the C-terminal valine. Ammonolysis has proved quite useful in the synthesis of oxytocin

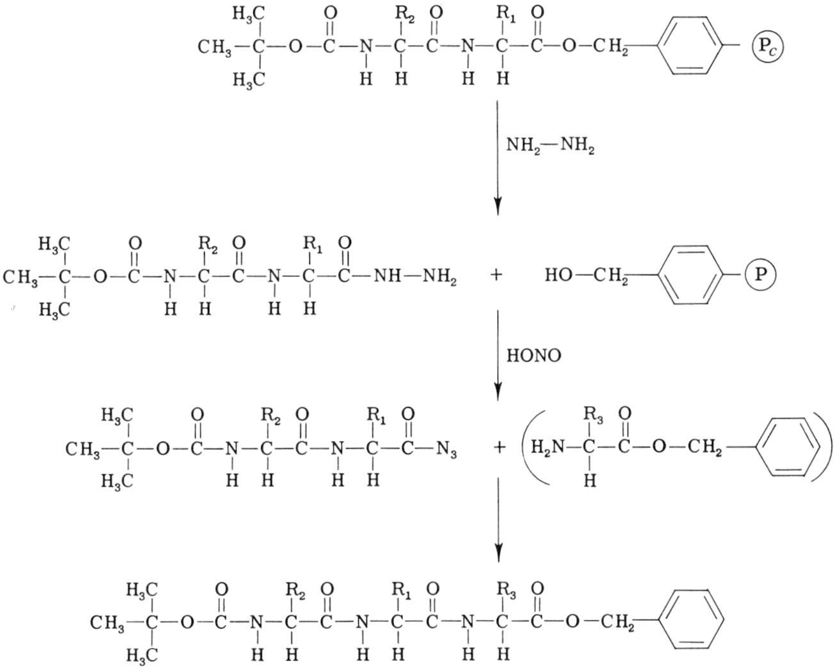

Fig. 4. Removal of protected peptides by hydrazinolysis and conversion to the azide for subsequent coupling.

(M. Manning, 1968; Beyerman *et al.*, 1968a; Inukai *et al.*, 1968; Bayer and Hagenmaier, 1968; Ives, 1968) and vasopressin (Meienhofer and Sano, 1968) by the solid phase method as well as in removing tryptophanyl peptides for enzymic hydrolysis (G. R. Marshall, 1968). Hydrazinolysis (Bodanszky and Sheehan, 1964; Kessler and Iselin, 1966; Ohno and Anfinsen, 1967; Stryer and Haugland, 1967) offers the advantage that the resulting peptide can be converted to the azide for further coupling reactions as shown in Fig. 4.

Although transesterification was first noted as a side reaction (Bodanszky and Sheehan, 1966), it has been made the basis of a method for removal of peptides from the resin. Other bases such as triethylamine (Lombardo and Piasio, 1968; Beyerman *et al.*, 1968b) and hydroxide forms of anion-exchange resins (Halpern *et al.*, 1968; Pereira *et al.*, 1969) have been used to catalyze the reaction. Transesterification with benzyl alcohol to give benzyl esters which could then be removed by hydrogenolysis is also possible (G. R. Marshall and Flanigan, 1969).

The original link (Merrifield, 1962, 1963) to the polymer support was either a nitrobenzyl or bromobenzyl ester. This was used because of the carbobenzoxy group which required HBr/HOAc for its removal. The increased stability to acidic cleavage procedures required the use of saponification as a cleavage method with the accompanying possibility of racemization. The polystyrene support was synthesized by either bromination or nitration of the chloromethyl derivative as shown in Eq. (5). Esterification by the normal procedure gave an amino acid

$$\text{Cl-CH}_2\text{-C}_6\text{H}_3(\text{Br})\text{-P} \xleftarrow{\text{Br}_2} \text{Cl-CH}_2\text{-C}_6\text{H}_4\text{-P} \xrightarrow{\text{HNO}_3} \text{Cl-CH}_2\text{-C}_6\text{H}_3(\text{NO}_2)\text{-P} \quad (5)$$

resin ester which was compatible with the use of the carbobenzoxy group for amino protection. Takashima *et al.* (1968) used nitrated resin for a recent synthesis of an oxytocin derivative in the expectation that it would be more susceptible to ammonolytic cleavage, although no clear advantage was noted.

The modified benzyl ester has also been used with a resitol-type support derived from the polymerization of phenol and formaldehyde (Losse *et al.*, 1968b) as shown in Fig. 5. The hydroxyls were blocked by reaction with diazomethane and the resulting polymer was chloromethylated. It was then esterified in ethyl acetate with a Boc amino acid. The coupling reaction was performed in

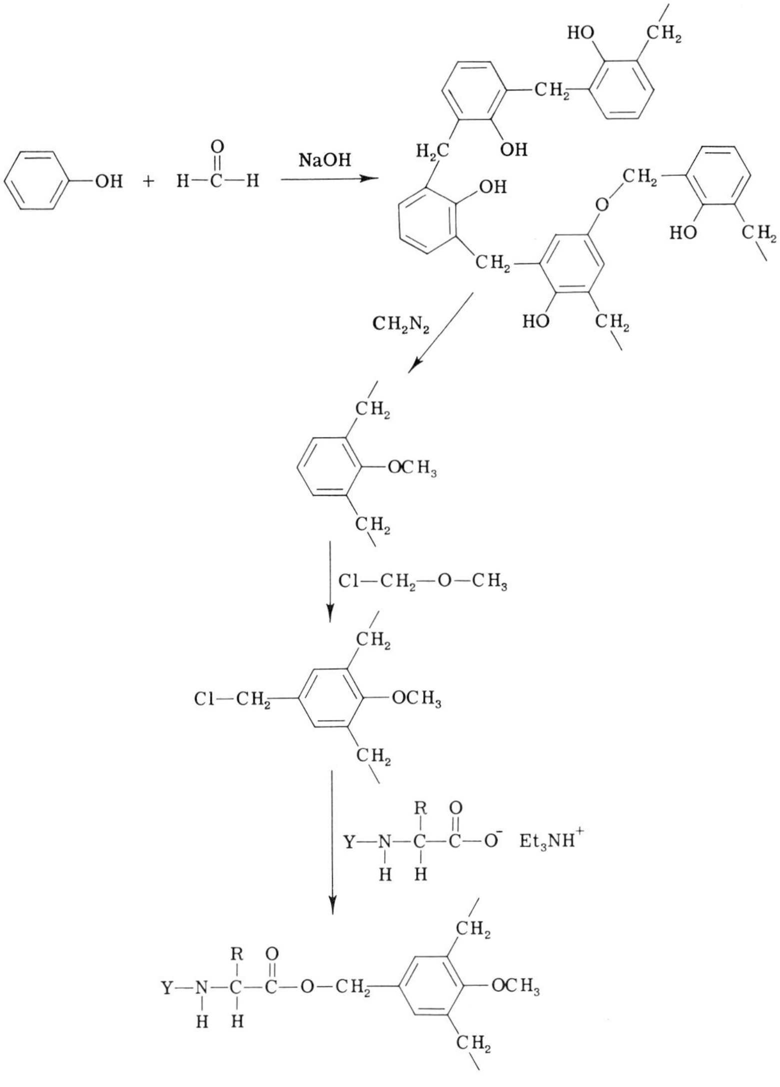

Fig. 5. Preparation of a resitol polymeric support to which the amino acid is attached by a modified benzyl ester bond (after Losse et al., 1968b).

dimethylformamide with DCCI. A penta- and heptapeptide were successfully synthesized on this polymer by Losse and co-workers. No advantages or disadvantages over the normal polystyrene support were noted. Similar polymers have been used with phenyl ester linkages for supports in solid phase peptide synthesis (Inukai et al., 1968; Flanigan and Marshall, 1970) as discussed below.

B. Phenyl Esters

Phenyl ester supports were introduced to solid phase synthesis by Inukai et al. (1968) in an ingenious synthesis of oxytocin in which the intrachain disulfide bridge was synthesized prior to cleavage of the peptide from the support. The polyphenol support was synthesized by reacting phenol and equimolar S-trioxane in bis(2-ethoxyethyl) ether in the presence of p-toluenesulfonic acid as shown in Eq. (6). Appropriately protected, i.e., Aoc, Boc, and Z, amino acids were esterified

$$\text{Ph-OH} + \text{trioxane} \longrightarrow \left[-CH_2- \underset{HO}{\text{Ph}} \right]_n \qquad (6)$$

by reaction with DCCI in dimethylformamide. Excess hydroxyl groups were protected by reaction with acetic anhydride. The protecting group could be removed by the appropriate hydrogen halide in a suitable solvent.

The coupling reaction appeared particularly sensitive to the nature of the solvent and coupling reagent used. Dimethylformamide was judged best and pyridine satisfactory, while chloroform, benzene, ethyl acetate, and dioxane were not satisfactory for the coupling reaction. With the DCCI or p-nitrophenyl ester methods, coupling yields were excellent; less satisfactory results were obtained with other active esters, azide, or mixed anhydride coupling.

Coupling of acyl peptides as well as protected amino acids was noted, although no experimental details were given. The peptide was removed from the support as the carboxylate by saponification with 1 N sodium hydroxide in aqueous methanol, as the amide by 25% ammonia in methanol, or as the hydrazide by hydrazine hydrate in DMF. Rearrangement of asparagine and glutamine residues (with ammonolysis) to the β- and γ-linkages was observed.

The advantages of the support were exercised further by providing spatial separation on the resin and, thereby, allowing for the preferential formation of intramolecular disulfide bonds in the synthesis of oxytocin as outlined in Fig. 6. The use of S-ethylmercapto-L-cysteine allowed the removal of the ethylmercaptan group while still on the support by disulfide exchange with thiophenol. The two free thiol groups of the cysteinyl residues were converted to the

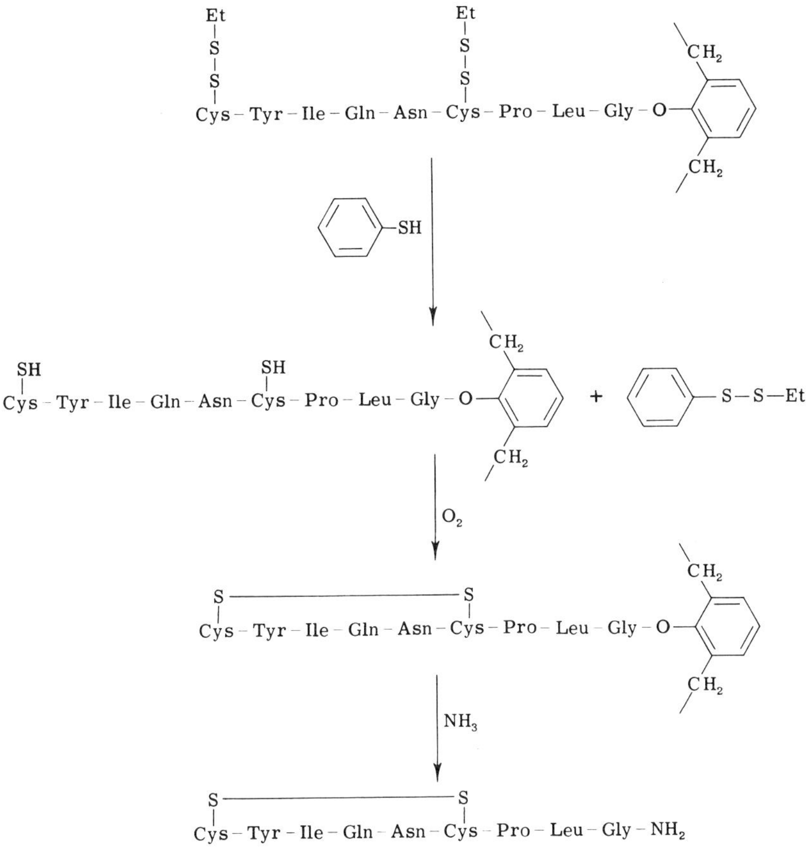

```
           Et                Et
           |                 |
           S                 S                    \
           |                 |                     CH₂
           S                 S                    /  \
           |                 |                   /    \
Cys – Tyr – Ile – Gln – Asn – Cys – Pro – Leu – Gly – O—⟨   ⟩
                                                       \    /
                                                        \  /
                                                         CH₂
                                                        /

                        ⟨   ⟩—SH
                          |
                          ▼

                                                      \
                                                       CH₂
   SH                  SH                             /  \
   |                   |                             /    \
Cys – Tyr – Ile – Gln – Asn – Cys – Pro – Leu – Gly – O—⟨   ⟩  +  ⟨   ⟩—S–S—Et
                                                       \    /
                                                        \  /
                                                         CH₂

                          | O₂
                          ▼
                                                      \
                                                       CH₂
   S ─────────────────── S                            /  \
   |                     |                           /    \
Cys – Tyr – Ile – Gln – Asn – Cys – Pro – Leu – Gly – O—⟨   ⟩
                                                       \    /
                                                        \  /
                                                         CH₂

                          | NH₃
                          ▼

   S ─────────────────── S
   |                     |
Cys – Tyr – Ile – Gln – Asn – Cys – Pro – Leu – Gly – NH₂
```

Fig. 6. Synthesis of oxytocin on a phenyl ester support according to Inukai *et al.* (1968).

intramolecular disulfide bridge by aeration and the peptide removed from the support by ammonolysis. A similar use of the polymer support for selective intrachain disulfide formation has been reported on a polystyrene support by Bondi *et al.* (1968).

Recently, the 4-(methylthio)phenyl ester of Johnson and Jacobs (1968) has been incorporated into a polymer support by Flanigan and Marshall (1970) and by D. L. Marshall and Liener (1969). This support offers the advantages of stable carboxyl protection for the synthesis of the peptide chain with possible activation by conversion to a 4-(methylsulfonyl)phenyl group by oxidation. The synthetic chain can then be removed by ammonolysis by amino acids, peptides, or other amino components, including its own amino terminus to form cyclic peptides.

Polymerization by Flanigan and Marshall (1970) of 4-(methylthio)phenol and formaldehyde, under basic conditions, gives a polymer of the resitol type shown in Eq. (7) similar to the polyphenol polymer of Inukai et al. (1968). An alternative method of incorporation by both groups (Flanigan and Marshall, 1970; D. L. Marshall and Liener, 1969) into a polymer form utilized the traditional 2% DVB-polystyrene beads shown in Fig. 7. D. L. Marshall and Liener (1969) reported briefly the synthesis of a linear, protected pentapeptide in which the

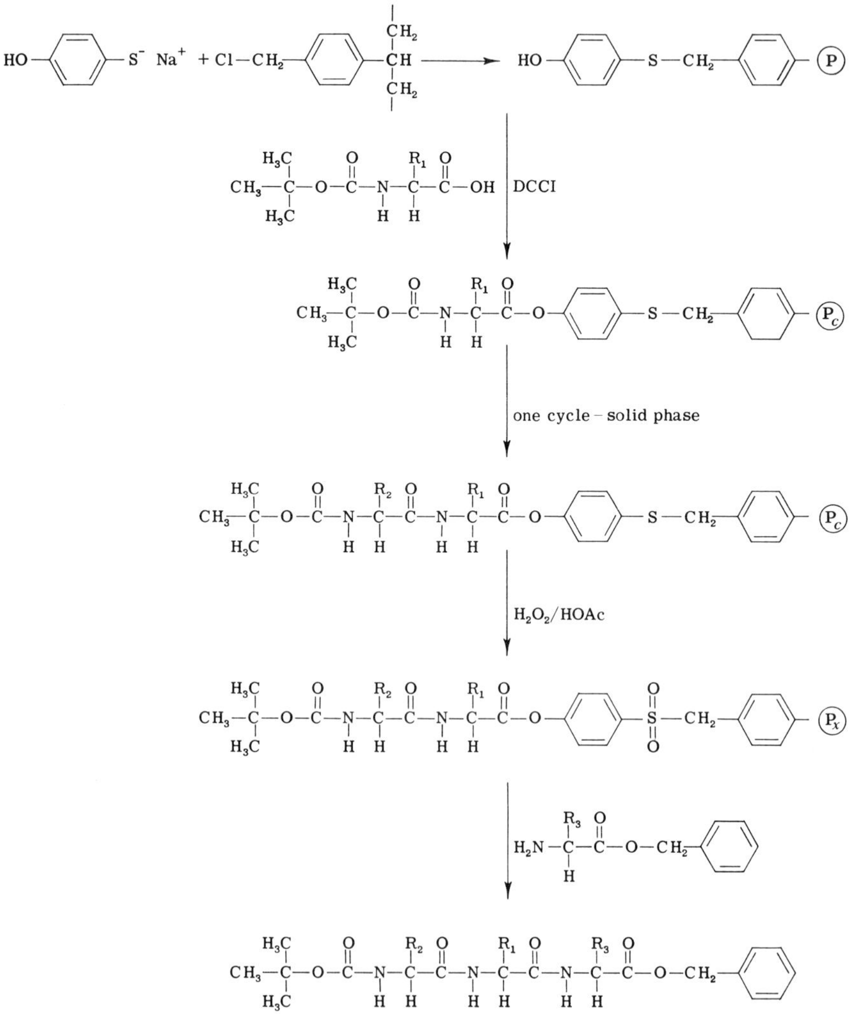

Fig. 7. Dual function support. Incorporation of 4-(methylthio)phenol group into polymeric support allows use both for carboxyl protection and activation. P_x, polymeric support used for carboxyl activation.

$$CH_3-S-\underset{}{\bigcirc}-OH + H-\overset{O}{\underset{}{C}}-H \longrightarrow \left[-CH_2-\underset{HO}{\bigcirc}\overset{S-CH_3}{} \right]_n \quad (7)$$

tetrapeptide was synthesized on the support and removed after activation by aminolysis with an amino acid ester. In a similar fashion, several small peptides were synthesized on both poly MTP supports by Flanigan and Marshall. Removal by intramolecular condensation from the activated supports gave good yields of cyclic di-, tetra-, and hexapeptides.

In contrast to the similar polymer of Inukai et al. (1968), the resitol-like polymer of 4-(methylthio)phenol swells in chloroform, methylene chloride, acetone, and tetrahydrofuran, as well as in dimethylformamide. It shrinks in alcohols and benzene. Protected amino acids were esterified primarily by the use of dicyclohexycarbodiimide in either dimethylformamide or tetrahydrofuran, or by the mixed anhydride procedure with ethyl chloroformate in tetrahydrofuran. Substitution in the range of 0.3–0.6 mmoles/g was common. Esterification with the coupling reagent (EEDQ) of Belleau and Malek (1968) in THF, with p-toluenesulfonic acid in benzene, and with thionyl chloride activation of the Li^+ form of the support according to Wieland and Birr (1966) have all been used successfully. Excess hydroxyl groups were blocked from further reaction either by acetylation with acetic anhydride or by methylation with diazomethane. The latter procedure is to be preferred if cleavage is to be by aminolysis with either an amino acid or peptide in order to prevent formation of N-acetyl derivatives. Only Boc derivatives have been used so far, although the stability of phenyl esters should allow the use of the wide variety of acid-labile groups available.

Again, in contrast with the otherwise similar polyphenol polymer, good yields were obtained by coupling both the dicyclohexylcarbodiimide in dimethylformamide or tetrahydrofuran, and with the mixed anhydride procedure in tetrahydrofuran. Cleavage of the peptide from the support was by conversion of the 4-(methylthio)phenyl ester to its activated derivative, the 4-(methylsulfonyl)phenyl ester, by oxidation. This was accomplished either by the use of 40% hydrogen peroxide in acetic acid or by m-chloroperbenzoic acid in dioxane. This gives a support which is chemically very similar to the activated ester support of Wieland and Birr (1967a) prepared from p,p'-dihydroxydiphenyl sulfone and formaldehyde. Cleavage from the activated support proceeds either by intramolecular aminolysis to give cyclic peptides, by saponification to give the free acid, by ammonolysis with ammonia or hydrazine to give the amide or hydrazide, or by reaction with an amino acid or peptide as the amino component in a coupling reaction.

The generality of such a support is limited by the oxidation conditions required, and would presently preclude the use of methionine, cysteine, and cystine in the growing peptide chain. Cleavage of the peptide from the unactivated polymer by methods similar to those used with the polyphenol support is quite feasible. In addition, sensitive amino acids can be handled by incorporating them as the amino component during the cleavage of the peptide from the activated support. The idea of changing the function of the support from carboxyl protection to carboxyl activation at will is attractive, and other, more elegant solutions will undoubtedly arise.

Another resitol-type support (see Section I,A) has been reported by Losse *et al.* (1968b). It was, however, chloromethylated and a benzyl ester to the support established for the carboxyl-terminal amino acid. It is interesting to note that very similar use of solvents to the normal polystyrene support, namely, ethyl acetate for esterification and DMF for coupling with DCCI, gave a successful synthesis of a heptapeptide and pentapeptide.

C. Alkyl Esters

Tesser and Ellenbroek (1967) have reported an alkyl alcohol link to a polymeric support. This support was distinguished by its sensitivity to base because of the presence of the β-sulfonyl ethyl ester linkage which can β-eliminate. The polymer, based on the carboxyl-protecting group of Hardy *et al.* (1968) and of Miller and Stirling (1968) was derived by the reaction of chloromethylated polymer with β-mercaptoethanol. The resulting sulfide was oxidized to the sulfone with perbenzoic acid, and the amino acid was esterified with DCCI as shown in Fig. 8. Removal of the esterified amino acid, carbobenzoxyphenylalanine, was demonstrated upon exposure to sodium methoxide in absolute methanol.

Peptides have been synthesized on the polymer of Tesser and Ellenbroek (1967) prior to oxidation to the sulfone (G. R. Marshall and Flanigan, 1969). After the synthesis was complete, the sulfide was oxidized to the sulfone and the peptide was removed by β-elimination. Esterification was accomplished with the use of *p*-toluenesulfonic acid in benzene. Deprotection of the Boc amino acids with 1 N HCl/HOAc and coupling with DCCI in dimethylformamide gave satisfactory peptide formation. Oxidation to the sulfone was accomplished according to Hardy, Rydon, and Thompson (1968) using hydrogen peroxide in acetone with ammonium molybdate as catalyst. Cleavage by β-elimination in dilute base gave the free peptide. It was shown that the normal neutralization procedure for solid phase synthesis, i.e., triethylamine, did not cause β-elimination. It was concluded that the oxidized support of Tesser and Ellenbroek would offer the advantage of being compatible with sulfur-containing amino acids and should also be compatible with the normal procedures for establishing the

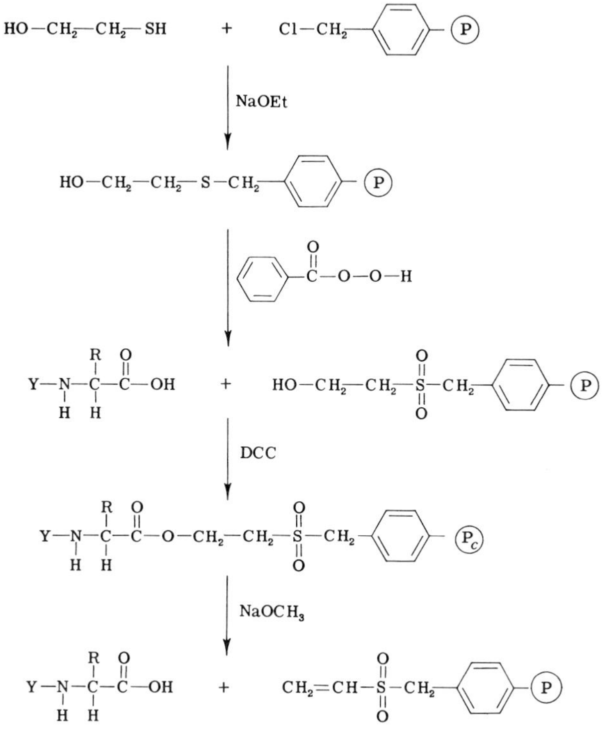

Fig. 8. Polymeric support described by Tesser and Ellenbroek (1967) in which cleavage of the bond to the support is by β-elimination.

peptide chain on polymeric supports. Work on the unoxidized support was terminated.

Chloromethylated polystyrene cross-linked with 2% divinylbenzene has been modified by reaction with ω-amino alcohols, followed by acetylation and saponification to give an alkyl alcohol. Another route to a similar product, again by Tilak and Hollinden (1968), was by a Friedel-Crafts reaction of the unmodified polymer with ω-halocarboxylic acid chlorides. The alkyl halide was converted to the alcohol by esterification with acetic acid, followed by saponification. These reactions are shown in Fig. 9. Carbobenzoxyamino acids were coupled by use of carbonyldiimidazole, and unreacted hydroxyls were blocked by acetylation. Deprotection with 1 N HBr in acetic acid and coupling with either mixed anhydride or dicyclohexycarbodiimide activation were found satisfactory. Peptides, including human insulin B-chain sequences 7–8, 15–20, and 24–30, were removed from the support either by saponification or by hydrazinolysis.

A support of exceptional potential is that of Weygand (1968) which contains a phenacyl halide. Styrene-divinylbenzene copolymer was either chloro- or bromo-acetylated through a Friedel-Crafts reaction as shown in Eq. (8). Weygand

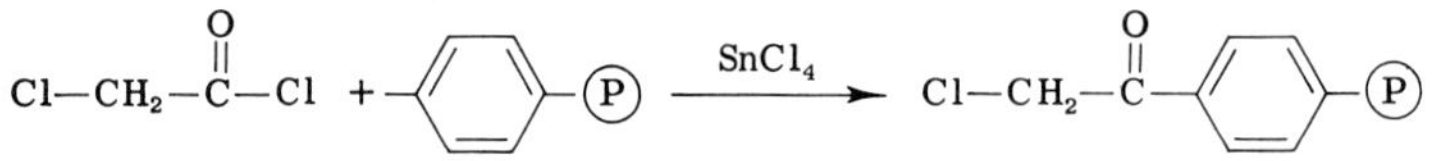

$$Cl-CH_2-\overset{O}{\underset{\|}{C}}-Cl + \text{⟨P⟩} \xrightarrow{SnCl_4} Cl-CH_2-\overset{O}{\underset{\|}{C}}-\text{⟨P⟩} \quad (8)$$

esterified a protected dipeptide, Boc-Ala-Ala-OH, to the polymer in the presence of the sterically hindered base, ethyldiisopropylamine, without the formation of quaternary nitrogen side products. N-Acyl peptide hydroxysuccinimide esters were reacted with the peptide on the support after its deprotection. The possibility of selective cleavage of the product with sodium thiophenol using protecting groups that are stable to this reagent was noted. This scheme is outlined in Fig. 10. The demonstration of the attachment of protected peptides to the polymer plus the potential of selective cleavage of the peptide from the support

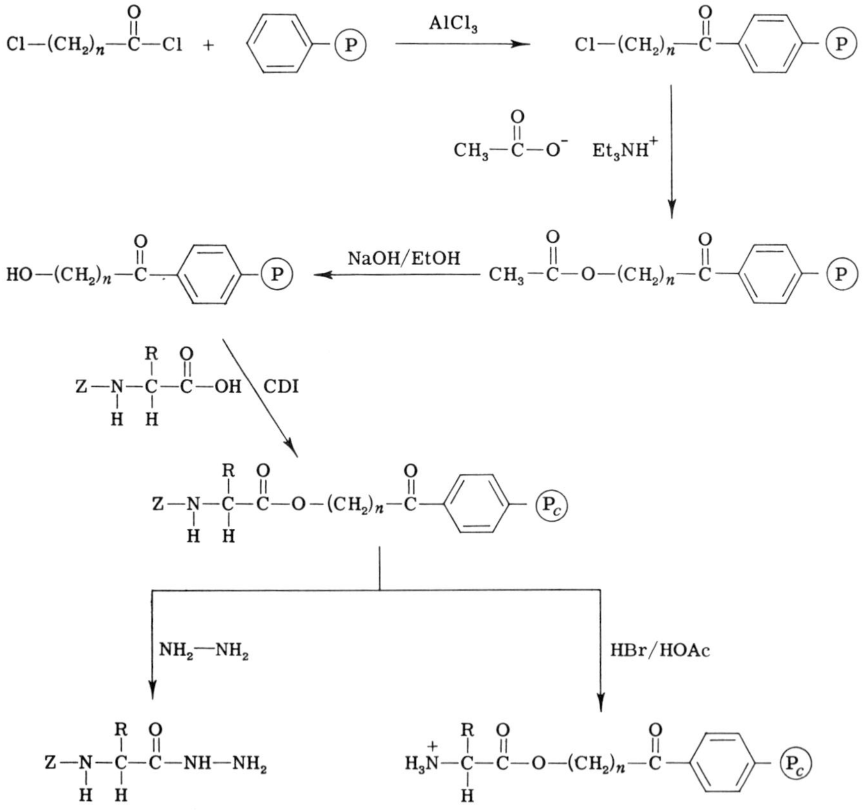

Fig. 9. Alkyl alcohol supports for use with carbobenzoxyamino acids as proposed by Tilak and Hollinden (1968). Z, carbobenzoxy (benzyloxycarbonyl) group.

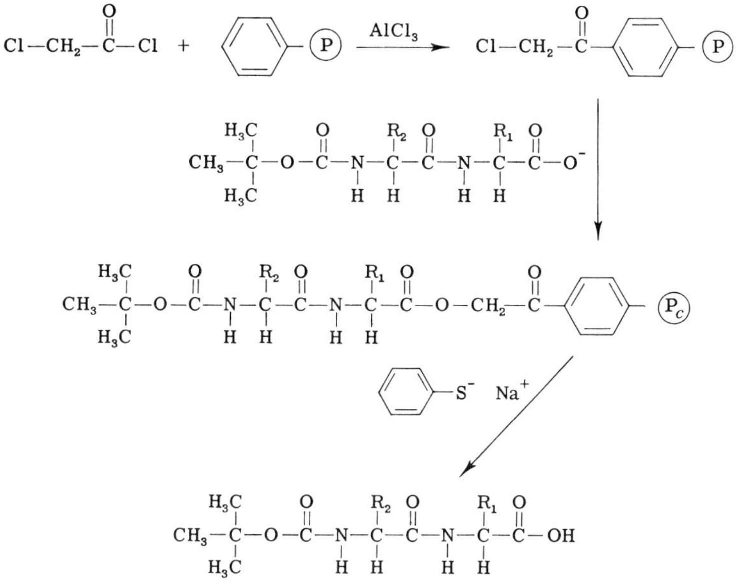

Fig. 10. Phenacyl support of Weygand (1968) with selective cleavage by thiophenol.

suggests several interesting possibilities. It should be possible, at least theoretically, to remove a growing peptide chain for characterization and purification, to restore its covalent bond to the support, and to continue the synthesis. This strategy, if it can be implemented, would appear to offer all the advantages of both solid phase and classical solution methods of peptide chemistry. The coupling of peptides rather than single amino acids offers the advantage of maximizing the effects of incomplete reactions which should simplify purification. As longer and longer peptides are synthesized, the problems of error sequences become greater. Peptide coupling with solid phase synthesis has been reported also by Sakakibara et al. (1968a), with DCCI activation by Inukai et al. (1968) and by Omenn and Anfinsen (1968), with activation by N-ethyl-5-phenyl isoxazolium-3-sulfonate (Woodward et al., 1961), by DCCI and hydroxysuccinimide (Weygand and Ragnarsson, 1966), and by the azide procedure.

The synthesis of an α-bromopropionyl polymeric derivative as well as the α-bromoacetyl polymer and their use have recently been reported (Mizoguchi et al., 1969). Aoc and Boc amino acids were esterified to the support and removed by saponification, ammonolysis, and hydrazinolysis. Boc-Leu-Leu-Tyr (Bzl)-OH was prepared in 36% yield from the Boc-O-benzyl-tyrosyloxyacetyl support and in 13% yield from the analogous propionyl support by saponification with 0.5 N NaOH/dioxane (1 : 2 by volume). A support incorporating an analog of the phthalimidomethyl-protecting group of Nefkens (1962) has been reported briefly by Wildi and Johnson (1968). An alternating copolymer of

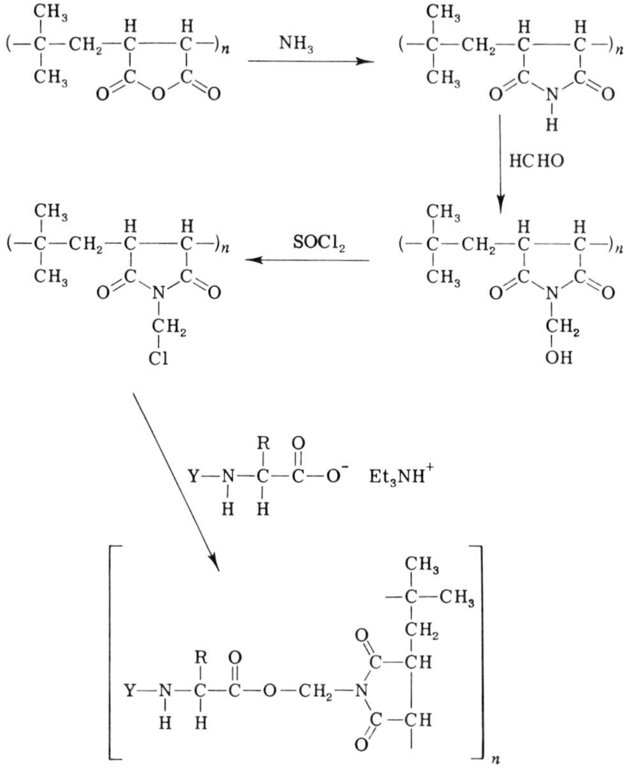

Fig. 11. Polymeric succinimidomethyl support derived from a copolymer of isobutylene and maleic anhydride (Wildi and Johnson, 1968).

isobutylene and maleic anhydride was converted to the imide with ammonia. Reaction with formaldehyde gave the hydroxymethylimide which was converted to the chloromethylimide by thionyl chloride. Esterification gave an amino acid protected by the polymeric succinimidomethyl group. The reactions are shown in Fig. 11. The lability of this group to acid resembles that of the phthalimidomethyl group and would require the use of the more labile amino-protecting groups such as the Nps, Bpoc, or Bip. These supports resemble the polymers which are used as activating agents, polymeric hydroxysuccinimide esters, which have been investigated by Wildi and Johnson (1968) as well as by Laufer and co-workers (1968) and by Patchornik and colleagues (1967).

A new general scheme has recently been developed by Wang and Merrifield (1969c). It was designed to give protected peptides of intermediate size for subsequent incorporation into larger polypeptides by fragment condensation methods. The scheme makes use of three classes of protecting groups with different degrees of sensitivity to acidic reagents. The α-amino protection is with the very acid-labile biphenylisopropyloxycarbonyl (Bpoc) group, the side-chain protection is based on the relatively stable benzyl group, and the carboxyl attachment to the resin support is through a t-alkyloxycarbonylhydrazide derivative of intermediate stability. The latter required the synthesis of a new resin derivative as shown in Fig. 12. Treatment of copolysytrene-divinylbenzene resin beads with methylvinylketone in liquid HF gave the polymer-linked methyl ketone which was converted to the tertiary alcohol by a Grignard reaction with methyl magnesium bromide. The alcohol was treated with phenylchloroformate and the phenyl ester was hydrazinolyzed to give the t-alkyloxycarbonylhydrazide resin support. A Bpoc amino acid was coupled to the acylhydrazide with DCCI in methylene chloride, and the protecting group was quantitatively removed by treatment with 0.5% TFA in methylene chloride. It was estimated that this

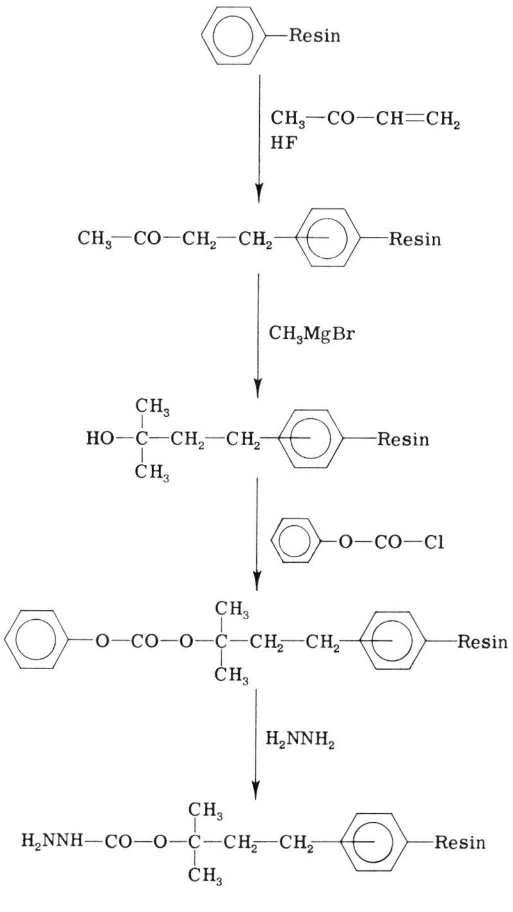

Fig. 12. Synthesis of a t-alkyloxycarbonylhydrazide resin.

treatment would give less than 6% loss of peptide chains from the resin during 40 synthetic cycles. Final cleavage from the resin was achieved with 50% TFA. To demonstrate the usefulness of this approach to a solid phase-fragment synthesis two protected peptide hydrazides were prepared: Z-Phe-Val-Ala-Leu-NHNH$_2$ and Z-Asp(Bzl)-Arg(NO$_2$)-Val-Tyr(Bzl)-NHNH$_2$.

D. Benzhydryl Polymeric Support

A new and potentially very exciting support is the polymeric benzhydryl ester of Southard and co-workers (1969). The resin form of the carboxyl-protecting group introduced by Hiskey and Adams (1966) and Stelakatas et al. (1966) was combined with a relatively new amino-protecting group, 1-benzoylisopropenyl (Bip), which has been used mainly in cysteinyl peptides by Hiskey and Southard (1966). The Bip-protecting group can be cleaved selectively in the presence of

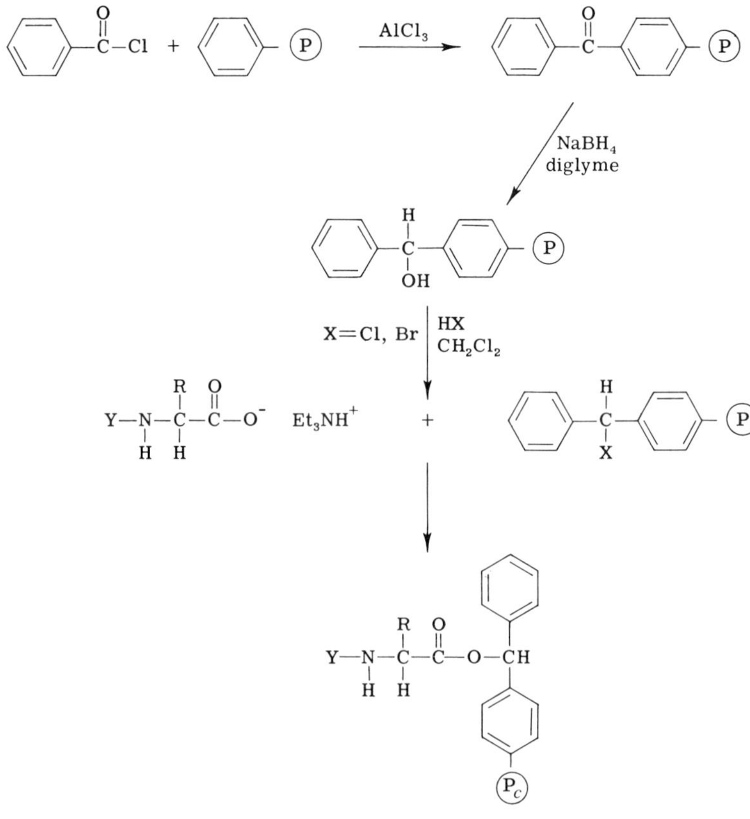

Fig. 13. Preparation of the polymeric benzhydryl ester according to Southard et al. (1969).

other acid-labile protecting groups by either 0.4 N aqueous HCl in tetrahydrofuran (THF) or 1 N p-toluenesulfonic acid monohydrate in THF. Cleavage of the peptide from the benzhydryl support is by treatment with 5–50% trifluoroacetic acid in chloroform for 30 minutes.

The benzhydryl support is prepared as shown in Fig. 13 by a Friedel-Crafts condensation of benzoyl chloride with cross-linked polystyrene. The resulting ketone is reduced with sodium borohydride and the hydroxyl replaced with a halogen by the appropriate hydrogen halide. The dicyclohexylammonium salt of the Bip-protected amino acid was esterified by refluxing in chloroform overnight. After exposure of the amino group by treatment with 0.4 N (aqueous) HCl-THF, the resulting hydrochloride was neutralized with triethylamine. The Bip amino acid was coupled as the dicyclohexylammonium salt by the addition of 1 mole of p-toluenesulfonic acid with the coupling agent, DCCI. Chromatographically pure Leu-Ala-Gly-Val was synthesized in 76% yield. It was quantitatively cleaved by aminopeptidase M digestion.

The 1-benzoylisopropenyl-protecting group is prepared by the condensation of benzoyl acetone with the amino acid as shown in Eq. (9), followed by

$$\text{Ph-C(=O)-CH}_2\text{-C(=O)-CH}_3 + \text{H}_2\text{N-CHR-C(=O)-OH} \longrightarrow \text{Ph-C(=O)-CH=C(CH}_3\text{)-NH-CHR-C(=O)-OH} \quad (9)$$

crystallization as the dicyclohexylammonium salt. The products are described as crystalline, high melting solids; readily soluble in ethanol, DMF, chloroform, or methylene chloride; and with a shelf stability at room temperature of at least 2 years.

Although there is no experimental evidence, the Bip group should be compatible with the t-alkyloxcarbonyl hydrazide of Wang and Merrifield (1969c). The benzhydryl support should be compatible with amino-protecting groups with lability similar to the Bip such as the Nps and Bpoc groups. It would be interesting if protected peptides could be easily attached to this support because the possibility of synthesis, cleavage, and purification, followed by reattachment, and further synthesis is very attractive.

E. Amide Support

The presence of numerous peptide hormones which have C-terminal amide groups, i.e., secretin, gastrin, thyrocalcitonin, etc., and some difficulty with ammonolysis of ester links (Bodanszky and Sheehan 1966) suggest attachment of the peptide through the amide function. This would also offer a selective advantage in that ammonolysis as a method of cleavage complicates the selection of side-chain protecting groups for aspartic and glutamic acid residues. Preliminary investigations (Pietta and Marshall, 1970) would suggest the usefulness of such an approach. Modified benzhydryl amine polymers have been synthesized by reaction of the benzhydryl bromide intermediate of Southard and co-workers (1969) with a saturated solution of ammonia in methylene chloride at 0° as shown in Eq. (10).

$$\text{Ph-CH(Br)-C}_6\text{H}_4\text{-(P)} \xrightarrow{\text{NH}_3} \text{Ph-CH(NH}_2\text{)-C}_6\text{H}_4\text{-(P)} \quad (10)$$

As an example of their use, benzhydryl amine polymer was reacted with Boc-Phe and DCCI. The remaining amino groups were acylated with acetic anhydride. After removal of the Boc group by 1 N HCl/HOAc and neutralization with triethylamine, Boc-β-benzyl-L-aspartic acid was coupled with DCCI. The dipeptide amide, H-Asp-Phe-NH$_2$, was obtained by treatment of an aliquot with HF. Two additional cycles resulted in the synthesis of gastrin tetrapeptide. This demonstrated that the benzhydryl amide support was stable to the routine procedures of solid phase peptide synthesis, and that the amide could be obtained by the use of HF. Modification of this type of resin as well as supports based on other amide links should offer a variety of polymers with varying degrees of lability for the synthesis of peptide amides.

III. POLYMERIC AMINO PROTECTION

Solid phase synthesis of a peptide chain can also be accomplished by attachment of the terminal amino group to a solid support and the stepwise extension of the chain at the carboxyl end. There are two general problems associated with this strategy which have limited its usefulness: one is racemization and the other is excessive loss of peptide chains because of side reactions. Since peptide bonds are formed by activation of the carboxyl component, this approach requires the continual activation at each step of the synthesis (after the first) of a *peptide* carboxyl in contrast to the activation of an *amino acid* carboxyl. The protection against racemization afforded by urethane amino-protecting groups is therefore lost, and most coupling methods will lead to partially racemic products. In addition, the activated component cannot be used in excess, and any incomplete reactions or side reactions will be reflected in a lowered yield and purity of product.

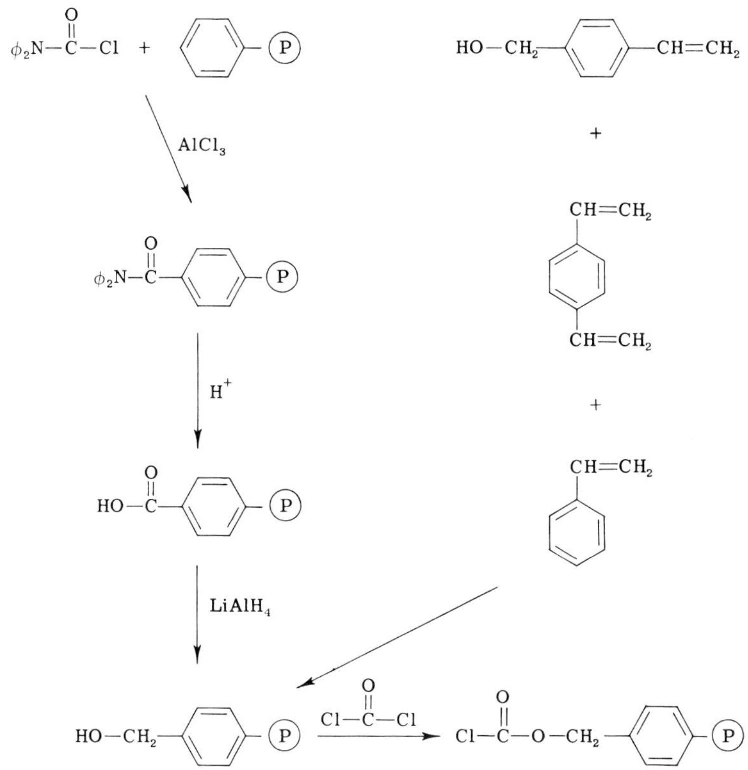

Fig. 14. Preparation of the original polymeric support used for amino protection, polymeric carbobenzoxy chloride (Letsinger and Kornet, 1963).

Of the entire gamut of amino-protecting groups that have been used in peptide chemistry (Schröder and Lübke listed over 20 in 1965), only one has been used for the linkage between the peptide chain and the solid support. Letsinger and co-workers (1964; Letsinger and Kornet, 1963) obtained a carbobenzoxy chloride-containing polymer through the copolymerization of vinyl benzyl alcohol with polystyrene and divinylbenzene followed by reaction with phosgene as well as by modifying preformed styrene-DVB with diphenyl carbamyl chloride followed by acid hydrolysis, by reduction with $LiAlH_4$, and by reaction with phosgene. These two routes are shown in Fig. 14. This polymer could then be reacted with an amino acid and the chain extended from the amino-terminal residue. For example, leucine ethyl ester was reacted, then saponified and coupled with glycine benzyl ester to give a dipeptide as shown in Fig. 15.

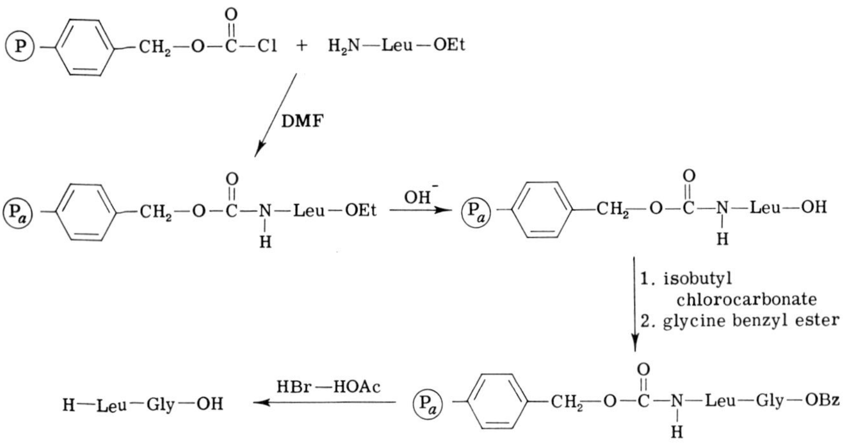

Fig. 15. Synthesis of leucylglycine on the polymeric carbobenzoxy support according to Letsinger and Kornet (1963). P_a, polymeric support used for amino protection.

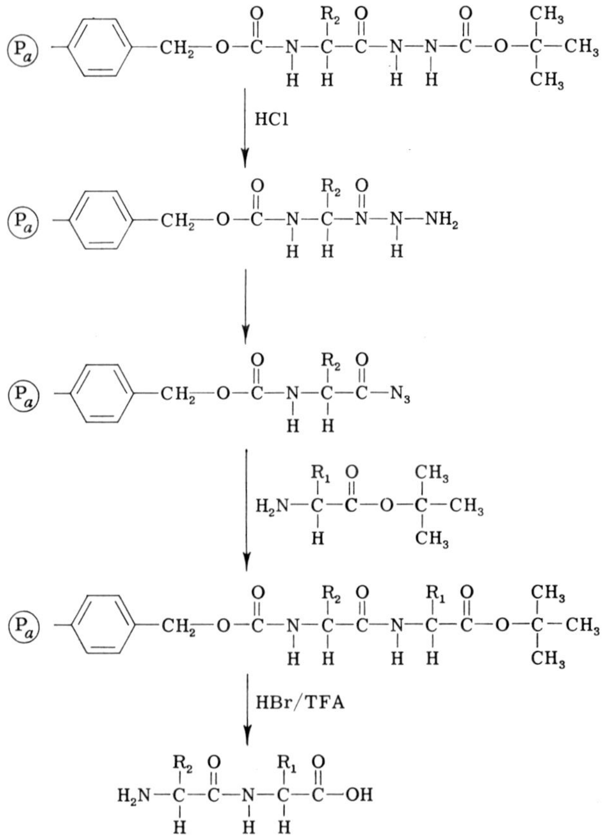

Fig. 16. Combination of polymeric amino protection with azide coupling through the use of Boc-hydrazide derivatives (Felix and Merrifield, 1970).

methylene chloride. The couplings were carried out with a mixed anhydride procedure. With these two modifications poly-γ-glutamyl folic acid was successfully synthesized.

Linkage of a growing peptide chain through a side chain group to the resin offers two general advantages. In certain instances use can be made of a particular side-chain functional group for a specific cleavage reaction. For example, linkage of the mercaptan group of cysteine to a support through a disulfide bond could provide a selective removal by disulfide interchange. In other instances a linkage to the polymeric support by a side chain near the middle of the peptide would make it possible to extend the chain either from the carboxyl or amino end, thereby facilitating preparation of a series of analogs with changes at either or both ends of the molecule as illustrated in Fig. 17. The major disadvantage of this approach, and the probable reason for the lack of its development, is the fact that any

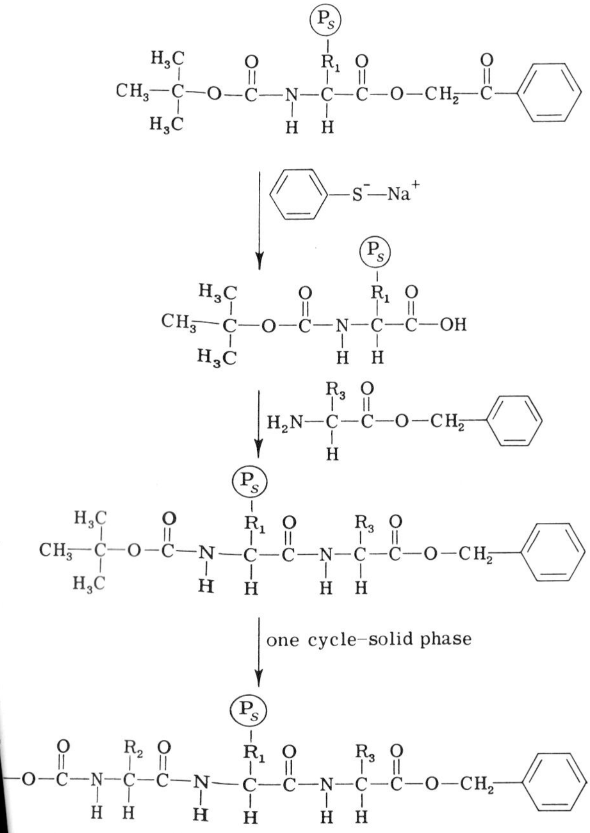

[Figure 17: scheme showing polymeric side-chain protection reaction sequence]

one cycle–solid phase

General scheme for polymeric side-chain protection which would allow peptide chain by additions to either end. P_s, polymeric support used for protection.

In a more recent study of this strategy, Felix and Merrifie[ld] [investi]gated the use of azide coupling, which was considered to be lar[gely racemization] free. They used the amino acids protected as the N'-Boc-hy[drazides.] The Curtius rearrangement was not observed and the decom[position] to the amide was avoided by coupling at low temperature[s. The] procedure is shown in Fig. 16. The cycle of extending th[e peptide con]sisted of (1) removal of the protecting group by treatment [with TFA] at 23° for 30 minutes, (2) conversion to the resin amino a[cid azide] with N-butylnitrite in THF at $-30°$, and (3) coupling [with the next] peptide Boc-protected hydrazide in THF using temper[atures of $-20°$] and 25°. For the last cycle of the synthesis, a t-butyl e[ster was substituted] for the Boc-hydrazide group to give the free acid upon [cleavage] with HBr-TFA. The tetrapeptide L-leucyl-L-alanyl[-... was syn]thesized in 30% yield by this stepwise procedure, a[nd also a] dipeptide, glycyl-L-valine t-butyl ester, was coupl[ed... It] was determined that preparation and coupling of res[in azides proceeded] in THF or DMF, but that dioxane, methylene c[hloride, and DMSO gave] poor results. It was concluded that this approach [has some] advantages, but that it should not be considered a[s superior to the] procedure using a carboxyl attachment to the re[sin.]

IV. POLYMERIC SIDE-CHAIN PROTECTI[ON]

It has been demonstrated recently that i[t is possible] to link the growing peptide chain to the [resin through a side chain.] Skylarov and Shaskova (1969) coupled a te[trapeptide, Boc-Val-] OMe, through the δ-amino group of orn[ithine to a chloromethyl] polymer (see Table III). The tetrapep[tide was prepared using] Nps-amino acids with DCCI-HOSu c[oupling. After Nps-] Orn(Z)-Leu-D-Phe-Gly-L-Val-Orn(pol[ymer)-OMe was syn]thesized, the methyl ester was saponi[fied and the azide] was prepared with DCCI. The Boc [group was removed and the peptide] allowed to cyclize. The cyclic peptid[e was cleaved from the resin] in acetic acid and was purified by i[on-exchange chromatography.]

A benzyl ester link to either the [γ-carboxyl] of glutamic acid would allow extens[ion of the peptide from the amino] or α-carboxyl, end. Krumdieck an[d Wilson attached Boc-glutamic] acid to a chloromethyl resin but [found that the benzyl ester was not] entirely stable to 1 N HCl/acet[ic acid. Thus, during removal of Boc] from the resin the deprotection [was accompanied by some loss...]

Fig. 17. Gener[al scheme for] extension of the pe[ptide chain from a] side-chain protecti[ng group.]

methodology would be specific for a particular amino acid. This would restrict the application to those peptides containing that amino acid; for example, with a disulfide link it would be applicable only to peptides containing cysteine or homologs of cysteine. With the proliferation of solid phase supports, however, it would not be too surprising to see more application of this approach in the near future.

V. POLYMERIC COUPLING REAGENTS

The use of polymeric activating agents was introduced into peptide chemistry almost simultaneously by Fridkin, Patchornik, and Katchalski (1965a) and by Wieland and Birr (1966). In this approach, the amino acid or peptide is reacted with a polymeric support which functions as the activating agent. In other words, the activating or active polymer is charged with either amino acid or peptide. Since the mechanism of activation of almost all the normally used coupling

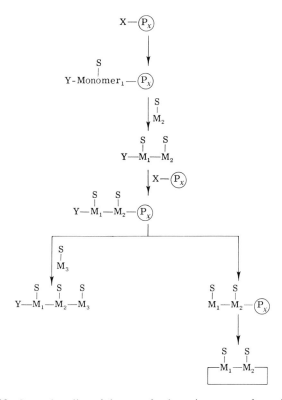

Fig. 18. General outline of the use of polymeric supports for activation.

reagents is through modification of the carboxyl group, the charging amino acid or peptide is usually an amino-protected derivative. Aminolysis of the charged polymer with the free amino group of either an amino acid or peptide results in the formation of a peptide bond and the release of the activated peptide from the support. A general outline of such a scheme is shown in Fig. 18. One particularly intriguing feature which was demonstrated by Fridkin et al. (1965b) was the use of the polymeric support to prevent intermolecular condensation during cyclization by spatial separation of the peptide chains as a result of their attachment to the polymer carrier. This resulted in an increased proportion of cyclic peptide when intramolecular aminolysis was used. The primary advantage of such an approach over classical procedures is that it allows the activated amino acid to be used in large excess with subsequent easy removal by filtration. In contradistinction to the normal solid phase procedure utilizing polymeric carboxyl protection, the product is readily available for further purification and characterization. One disadvantage of this approach is the possibility of racemization when charging the activating polymer with a peptide. This becomes of particular concern when one is attempting to exploit the polymer for the synthesis of cyclic peptides. Successful use of the technique in which the polymer supplies a source of activated amino acid has resulted in the synthesis of bradykinin in an overall yield of 12%. One theoretical disadvantage, which has not yet been demonstrated, is the problem of diffusion when one is attempting to couple a large peptide with a large activated amino acid polymer.

A. Polymeric Active Esters

1. *Nitrophenyl Esters*

The polymeric nitrophenyl ester was introduced by Fridkin and co-workers (1965a) in the form of a poly-4-hydroxy-3-nitrostyrene cross-linked with 4% divinylbenzene. Acyl amino acids and peptides were coupled to this support by use of DCCI, mixed anhydride, azide, and nitrophenyl ester methods. The polymeric nitrophenyl ester, or charged polymer, could then be reacted with a carboxyl-protected amino acid, cleaving the product from the support. Blocked glutathione was synthesized in this manner as shown in Fig. 19 in 95% yield. An overall yield of 12% of purified biologically fully active bradykinin was obtained by Fridkin et al. (1968a) using this procedure. Steric problems in the coupling of Leu, Ile, and Val which result in low yield have been reported recently (Fridkin, 1971). Preliminary investigation of a derivative of the chloromethyl polystyrene polymer resulting from the reaction with p-nitro-o-catechol was also reported in an attempt to find a support with suitable physical properties.

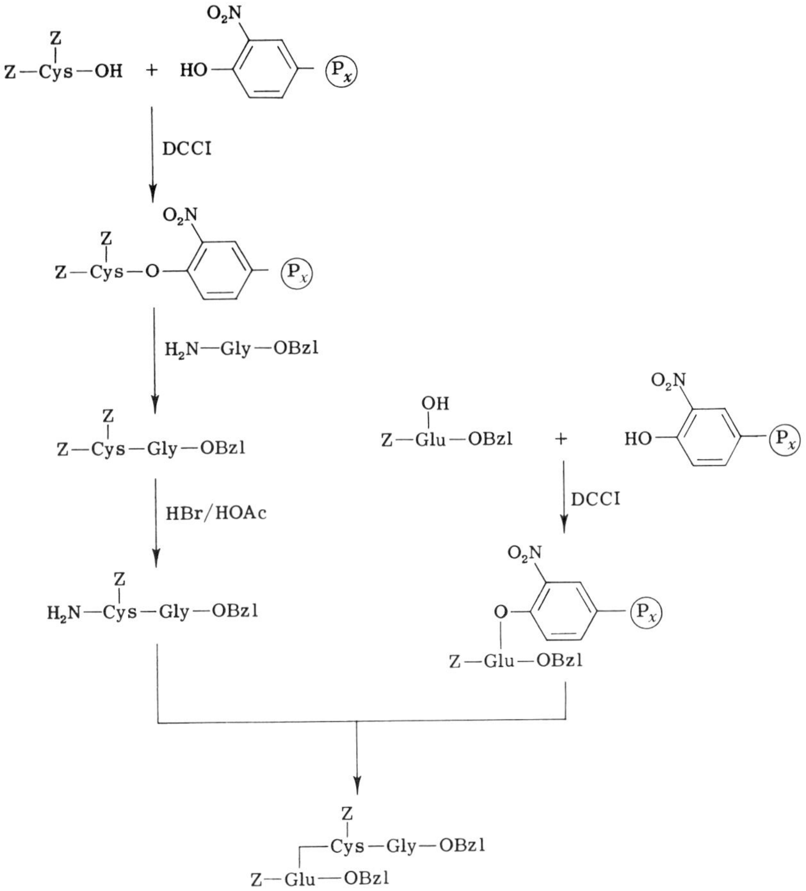

Fig. 19. Synthesis of blocked glutathione using polymeric activation (Fridkin *et al.*, 1966).

Patchornik *et al.* (1967) have also reported the use of nitrated resitol-type supports as well as branched copolymers of DL-lysine and 3-nitro-L-tyrosine as a basis for polymeric nitrophenyl esters although without detail. This group (Fridkin *et al.*, 1965b) of investigators realized the potential benefits of the spatial separation imposed on the peptides by the polymeric support and utilized it to provide a synthetic route to cyclic peptides without the normal intermolecular condensations which accompany cyclization, even at high dilution. Internal aminolysis by the amino end of the peptide chain after deblocking and neutralization leads to a cleavage of the cyclic peptide derivative. Several cyclic di- and tetrapeptides have been prepared in this fashion.

2. Sulfonylphenyl Esters

Wieland and Birr (1966, 1967a) prepared a copolymer of p,p'-dihydroxydiphenyl sulfone and formaldehyde. This polymer was then charged with a protected amino acid by either mixed anhydride or DCCI activation or by activation of the hydroxyl group with thionyl chloride as shown in Eq. (11). An

$$\text{Y-NH-CHR-COOH} + \text{HO-Ar-SO}_2\text{-Ar-OH} \xrightarrow{\text{SOCl}_2} \text{Y-NH-CHR-CO-O-Ar-SO}_2\text{-Ar-O-CO-CHR-NH-Y} \tag{11}$$

attempt was made to incorporate this polymer into a continuous process, and a photolabile amino protecting group, the 3,5-dimethoxybenzyloxycarbonyl group, was prepared to aid in the development. A tetrapeptide ester was synthesized in very good yield.

The 4-(methysulfonyl)phenyl ester-activating group of Johnson and Jacobs (1968) has been incorporated into a polymeric support by two groups recently. In both cases, the active ester support resulted from oxidation of a precursor, support, 4-(methylthio)phenyl ester, which was used for the establishment of peptide chains by the conventional protocol (see Section I,C). D. L. Marshall and Liener (1969) modified the chloromethyl DVB-polystyrene support by the addition of p-mercaptophenol as did Flanigan and Marshall (1970), who also used a polymer similar to that of Wieland and Birr (1966) prepared from 4-(methylthio)phenol and formaldehyde. This "dual function" support appears particularly useful for the preparation of cyclic and protected peptides.

3. Hydroxysuccinimide Esters

The use of this polymeric activating agent has been reported by Patchornik et al. (1967), Wildi and Johnson (1968), and Laufer et al. (1968). The synthesis of 17 crystalline peptides of sizes up to an octapeptide was reported by Laufer *et al.* (1968), using Boc-amino acids bound to copoly(ethylene-*N*-hydroxymaleimide). In this case the hydroxysuccinimide polymer was formed by the reaction of hydroxylamine with copoly(ethylenemaleic anhydride). The amino acid to be activated was coupled to the polymer with either DCCI or mixed anhydride activation. The reaction scheme is shown in Fig. 20.

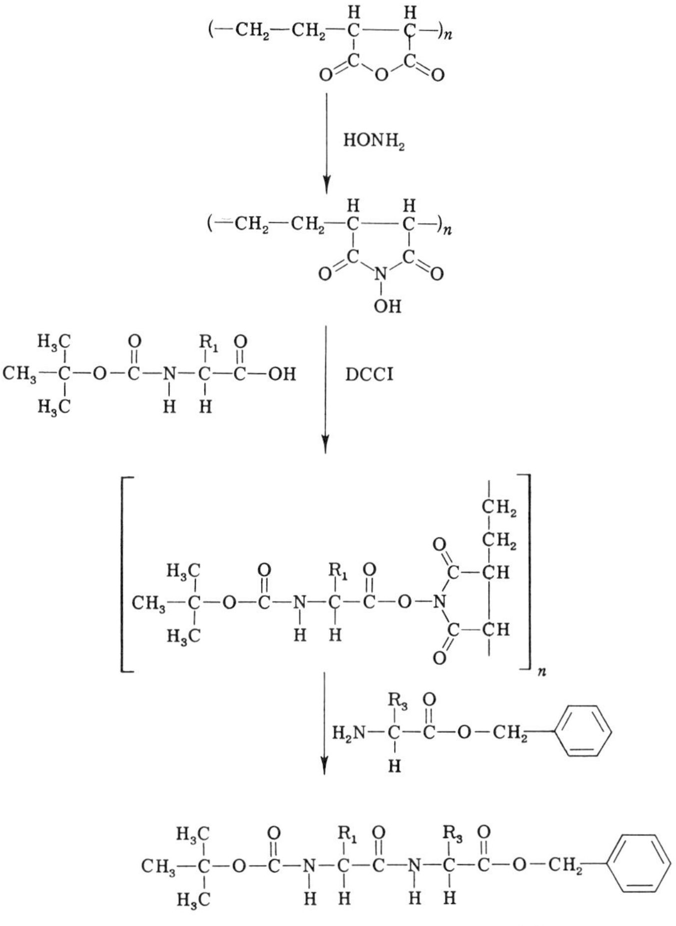

Fig. 20. Preparation and use of polymeric hydroxysuccinimide esters.

4. Quinaldine Esters

Manecke and Haake (1968) polymerized 5-vinyl-8-benzyloxyquinaldine with 20 mole % divinylbenzene as shown in Eq. (12). The benzyl group was removed

$$\text{(scheme 12)} \tag{12}$$

by hydrolysis with concentrated HCl, and the resulting hydroxyl group esterified with Z-DL-Ala-OH. Aminolysis with glycine ethyl ester gave the protected dipeptide ester with good yield and purity. An analogous procedure was performed in solution for comparison. The formation of the peptide on the polymeric support was ascertained by IR spectroscopy.

5. *Thiophenyl Esters*

The use of 4-mercaptostyrene as an insoluble active thioester was mentioned by Patchornik *et al.* (1967), but no experimental details were given.

B. Polymeric Carbodiimide

Polymeric hexamethylenecarbodiimide [Eq. (13)] has been used in peptide

$$[-CH_2-CH_2-CH_2-CH_2-CH_2-CH_2-N=C=N-]_n \tag{13}$$

synthesis by Wolman *et al.* (1967). The carboxyl component and amino component are added to an excess of the insoluble reagent. The unused polymer is separated after the reaction is completed. Rearrangement to acyl urea and possible racemization would appear to be limitations on the applicability of this procedure. Fridkin (1969) has reported briefly on several carbodiimide derivatives of the usual polystyrene support.

VI. EFFECTS OF PHYSICAL PROPERTIES OF POLYMERIC SUPPORTS

Although a variety of chemical links to the polymeric support have been investigated, the role of the physical nature of the support and its effects on the synthetic process have been sadly neglected. There is mounting evidence that this plays an important role in the practicality of solid phase synthesis. Considering the enormous polymer technology which exists, one can expect considerable investigation of the role of the chemical and physical nature of the polymer in solid phase synthesis. Table II lists the polymeric supports which have been reported and various properties such as solubility, polymeric role, physical form, cross-linking, covalent link, and polymeric class.

The earliest experiments were with cellulose, polyvinyl alcohol, and polymethylmethacrylate (Merrifield, 1963) which were not satisfactory under the rather harsh conditions. The first successful solid support and the most widely used polymer is the copolymer of styrene and divinylbenzene (Merrifield, 1962). Almost all of the polymeric modifications to date are based on a styrene polymer with variation in cross-linking, physical form, and covalent link to the monomer. Merrifield (1963) examined suspension polymers of polystyrene-DVB, prepared by the Dow Chemical Company, of various cross-linking ratios. Since the degree of cross-linking determines the solubility, extent of swelling, effective pore size, and mechanical stability of a polymer, the effect on the applicability to peptide synthesis is direct and can be dramatic. Originally, 1% cross-linking resulted in a fragmentation during synthesis and difficult filtration; although under modified conditions, the 1% DVB support proved not only stable but gave superior results (Merrifield, 1967). More highly cross-linked supports such as 8% and 16% DVB suffered from slower and less complete reactions, probably due to hindered diffusion.

It is conceivable that the reactions in a swollen polymer resemble those in a dilute gel since the rates are only slightly slower than in solution. One potential difficulty, which has not been overstressed (G. R. Marshall, 1968), is the change in solvation of the polymer as the growing peptide chain is extended. Normal substitution is 0.1–0.5 mmoles amino acid/g of polymer. After the addition of eight residues, the polymer will have increased its weight by approximately one-third. Over 30% of the polymer is now peptide, and its properties can be expected to reflect this dramatic change. This may, in fact, be part of the explanation of the sequence-dependent problems in solid phase synthesis; sequence-dependent phenomena plague both solid phase and classical peptide chemists. Just at the insolubility of protected peptide chains present difficulties with solution methods, the effect of adding a substantial percentage of insoluble residues to a polymer cannot be overlooked. As the solvation of the polymer is decreased, the accessibility of the heterogeneous population of sites can be expected to change, perhaps dramatically.

One of the most interesting and best documented cases of a dramatic change is the synthesis of angiotensinylbradykinin by Merrifield (1967). Protected bradykinin was synthesized normally (Merrifield, 1964a,b) and then the procedure of G. R. Marshall and Merrifield (1956) for the synthesis of [Ile^5]-angiotensin II was followed by adding eight more amino acids to the amino terminal of bradykinin.

TABLE II

	Polymer	Polymeric role	Physical form	Cross-linkage
1.	Styrene-DVB	Carboxyl prot.	Beads	1%
2.	Styrene-DVB	Carboxyl prot.	Beads	2%
3.	Styrene-DVB	Carboxyl prot.	Beads	4%
4.	Styrene-DVB	Carboxyl prot.	Beads	8%
5.	Styrene-DVB	Carboxyl prot.	Beads	16%
6.	Styrene-DVB	Carboxyl prot.	Macroreticular	High
7.	Styrene-DVB	Carboxyl prot.	Macroporous	High
8.	Stryene-DVB	Carboxyl prot.	Beads	2%
9.	Styrene-DVB	Carboxyl prot.	Beads	2%
10.	Styrene-DVB	Carboxyl prot.	Beads	2%
11.	Styrene-DVB	Carboxyl prot.	Beads	2%
12.	Styrene-DVB	Carboxyl prot.	Beads	2%
13.	Styrene-DVB	Amino prot.	Popcorn	0.2%
14.	Styrene	Carboxyl prot.	Amorphous	0%
15.	Phenol-formaldehyde	Carboxyl prot.	Amorphous	+
16.	Phenol-S-trioxane	Carboxyl prot.	Amorphous	+
17.	Methylthiophenol-formaldehyde	Dual	Amorphous	+
18.	Styrene-DVB	Dual	Beads	2%
19.	Styrene-DVB	Carboxyl prot.	Beads	2%
20.	Styrene-DVB	Carboxyl prot.	Beads	2%
21.	Styrene-DVB	Carboxyl prot.	Beads	2%
22.	Stryene-DVB	Carboxyl prot.	Beads	2%
23.	Styrene-DVB	Amino prot.	Beads	2%
24.	Styrene-DVB	Activation	Beads	4%
25.	Phenol-formaldehyde	Activation	Amorphous	+
26.	Phenol-formaldehyde	Activation	Amorphous	+
27.	Ethylene-maleimide	Activation	Amorphous	0%
28.	Quinaldine-DVB	Activation	Amorphous	20%
29.	Styrene-DVB	Carboxyl prot.	Beads	4%
30.	Methyl methacrylate-DVB	Carboxyl prot.	Beads	+
31.	Vinyl alcohol	Carboxyl prot.	–	0%
32.	Cellulose	Carboxyl prot.	–	+
33.	Dextran	Carboxyl prot.	Beads	+
34.	Isobutylene-maleimide	Carboxyl prot.	Amorphous	0%
35.	Styrene	Carboxyl prot.	Pellicular (coated glass)	–
36.	Brush	Carboxyl prot.	Glass beads	–
37.	Styrene graft	Carboxyl prot.	Styrene on Teflon	–

Properties of Polymeric Supports

Covalent link	Organic solubility	Polymeric class	Reference
Benzyl ester	I	Polystyrene	Merrifield (1963, 1967); Gutte and Merrifield (1969)
Benzyl ester	I	Polystyrene	Merrifield (1963, 1967, 1969a)
Benzyl ester	I	Polystryene	Merrifield (1963)
Benzyl ester	I	Polystryene	Merrifield (1963)
Benzyl ester	I	Polystyrene	Merrifield (1963)
Benzyl ester	I	Polystyrene	Merrifield (1965a)
Benzyl ester	I	Polystyrene	Merrifield (1965a)
Nitrobenzyl ester	I	Polystyrene	Merrifield (1963)
Bromobenzyl ester	I	Polystyrene	Merrifield (1963)
Sulfonylethyl ester	I	Polystyrene	Tesser and Ellenbroek (1967)
Alkyl ester	I	Polystyrene	Tilak and Hollinden (1968)
Phenacyl ester	I	Polystyrene	Weygand (1968); Mizoguchi et al. (1969)
Benzyloxycarbonyl	I	Polystyrene	Letsinger and Kornet (1963); Letsinger et al. (1964)
Benzyl ester	S	Polystyrene	Shemyakin et al. (1965); Garson (1968); Green and Garson (1969)
Benzyl ester	I	Resitol	Losse et al. (1968b)
Phenyl ester	I	Resitol	Inukai et al. (1968)
Methylthiophenyl ester	I	Resitol	Flanigan and Marshall (1970)
Methylthiophenyl ester	I	Polystyrene	Flanigan and Marshall (1970); Marshall and Lierner (1969)
t-Butyl ester	I	Polystyrene	Wang and Merrifield (1969c)
t-Butyloxycarbonyl hydrazide	I	Polystyrene	Wang and Merrifield (1969c)
Benzhydryl ester	I	Polystyrene	Southard et al. (1969)
Benzhydryl amide	I	Polystyrene	Pietta and Marshall (1960)
Benzyloxycarbonyl	I	Polystyrene	Felix and Merrifield (1970)
Nitrophenyl ester	I	Polystrene	Fridkin et al. (1965a, b; 1966; 1968a, b)
Nitrophenyl ester	I	Resitol	Patchornik et al. (1967)
Sulfonylphenyl ester	I	Resitol	Wieland and Birr (1966)
Hydroxysuccinimide ester	I	Copolyethylene-maleimide	Laufer et al. (1968); Patchornik et al. (1967); Wildi and Johnson (1968)
Hydroxyquinaldine ester	I	Polyvinylquinaldine	Manecke and Haake (1968)
Aminophenyl ester	I	Polystyrene	Merrifield (1963)
Aminophenyl ester	I	Polymethacrylate	Merrifield (1963)
Alkyl ester	?	Polyvinyl alcohol	Merrifield (1963)
Alkyl ester	I	Polysaccharide	Merrifield (1963)
Aminobenzyl ester	I	Polysaccharide	Merrifield (1969); Vlasov and Bilibin (1969)
Hydroxymethyl-succinimide ester	I	Copolyethylene-maleimide	Wildi and Johnson (1968)
Benzyl ester	I	Polystyrene	Bayer et al. (1970a)
Benzyl ester	I	Alkyl silicate	Bayer et al. (1970b)
Benzyl ester	I	Polystyrene-Teflon graft	Tregear et al. (1967)

The addition of residue number 12, im-benzyl histidine, was only accomplished in 20% yield, even though the identical histidylprolyl bond had been synthesized in angiotensin with no difficulty. The changes in procedure, after several attempts, which led to a successful synthesis are outlined in Fig. 21. Note

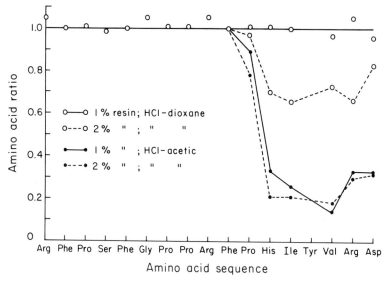

Fig. 21. Effect of solvent and resin cross-linking on the synthesis of angiotensinyl-bradykinin. The synthesis proceeds from left to right.

that the change from 2% DVB cross-linking to 1% did not, in itself, have any effect. The level of available sites dropped to approximately 20% and remained there. The substitution of 4 N HCl/dioxane for the 1 N HCl/HOAc in the deprotection step increased the number of available sites to approximately 70%. Using the combined charge, both 1% DVB and HCl/dioxane, gave a successful synthesis. Decreased solvation as a function of peptide structure could lead to a decrease of accessible areas of the polymer for critical reactions such as deprotection or coupling. Although not reported, the change in deprotection solvent was also accompanied by a change from DMF to $CHCl_3$ for the neutralization step. In both cases, the change was to a solvent with lower dielectric constant.

The fact that the reactive sites are distributed in the normal case throughout the gel matrix was demonstrated by autoradiography (Merrifield and Littau, 1968) as shown in Fig. 22. The sites have varying reactivity, however, reflecting heterogeneity at the molecular level. This has been suggested by preliminary rate measurements by Gut and Rudinger (1968). The measured rate constants

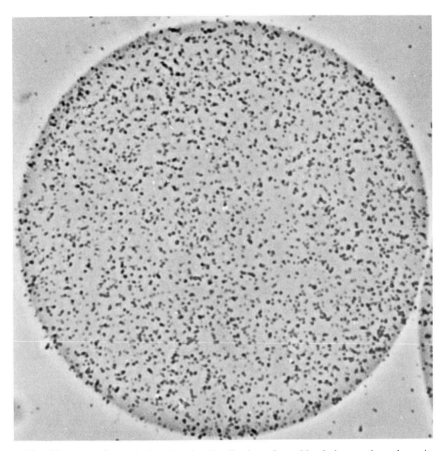

Fig. 22. Autoradiograph showing the distribution of peptide chains on the polymeric support. From Merrifield (1969a).

decreased as the coupling reaction proceeded indicating kinetically heterogeneous sites, although this conclusion may have to be modified once more refined measurements are available. Another interesting observation was an apparent selective uptake of solvent with exclusion of the solute by the resin. Gut and Rudinger (1968) conclude that " It is evident that distribution phenomena must be considered in solid phase synthesis." This is not an unknown occurrence; another example would be the use of water-soluble polymers to form an aqueous two-phase system for countercurrent distribution (Albertsson, 1960).

An alternative to making sites available by swelling the support to a state of complete solvation is to increase the surface area of the particle so that diffusion into the gel matrix is not an important factor. Rigid copolymers, both macroporous

(Millar *et al.*, 1963) and macroreticular (Kunin *et al.*, 1962), containing large pores fixed by a high degree of cross-linking with an effective surface area several hundred times that of suspension polymers have been preliminarily investigated (Merrifield, 1965a). A decreasing extent of reaction was observed as the peptide chain was elongated (Merrifield, 1969a).

Low density, popcorn, polymers of styrene and divinylbenzene have been examined by Letsinger and co-workers (1964; Letsinger and Kornet, 1963; Letsinger and Jerina, 1967). These materials appear to be freely penetrated by small molecules even though they do not swell in the solvents used, and are characterized by an extremely low (0.1–0.5%) cross-linking. Although functional groups appeared completely accessible in suitable solvents such as benzene, pyridine, or dimethylformamide, limited accessibility was apparent in hydroxylic solvents such as water or methanol. Letsinger and Jerina (1967) have compared popcorn and the normal suspension polymer having esterified benzoic groups with solution benzoate esters. Aminolysis of the *p*-nitrophenyl esters gave approximately equivalent rates with the two polymers; only 20% that in solution. Transesterification with benzyl alcohol gave a similar relation when catalyzed by *N*-methylimidazole. Catalysis by polymeric *N*-vinylimidazole showed a much slower reaction for the polymeric esters. The conclusion that large reagent molecules may present difficulties with these supports is equally applicable to the problems of coupling peptides to amino acid-charged, polymeric active ester supports (see Section IV).

Bayer (1971) has reported briefly on the use of pellicular (Horvath *et al.*, 1967) supports in solid phase synthesis and has suggested the use of brush (Halasz and Sebastian, 1969) polymers. The pellicular supports appeared to alleviate some deletions noted during the synthesis of model peptides with the normal support, but were deficient in their physical properties. Both are characterized by having the sites limited to near the surface, thus minimizing diffusion. Supports with similar properties have been made by graft polymerization (Tregear *et al.*, 1967) onto a dense polymeric interior.

Shemyakin and co-workers (1965, Ovchinnikov *et al.*, 1969) attempted to minimize the problems of diffusion by the use of soluble polymeric supports. Most of the work has used an emulsion polystyrene with average molecular weight of 200,000 which is soluble in organic solvents. Precipitation by pouring the reaction mixture into water was used for purification. One problem which is similar to that discussed above is the change in solubility properties with growth in the polypeptide chain. Garson (1968) and Green and Garson (1969) noted considerable changes in the physical state of the aqueous precipitate depending on the peptide chain, which made filtration difficult. Garson (1968) was also unable to cleave the peptide from the support by catalytic hydrogenation, probably due to steric interference from the polymeric support, which had already been shown for the solid support.

Other polymers based on isobutylene-maleic anhydride such as the methylmaleimide support of Wildi and Johnson (1968) and the active ester supports or on resitol-type supports such as that of Inukai et al. (1968) have not been characterized very thoroughly with regard to their physical properties.

The use of more hydrophilic supports such as polyamino acids (Patchornik et al., 1967), modified polysaccharides (Merrifield, 1969a), and the maleic anhydride supports (Wildi and Johnson, 1968) might offer more flexibility with regard to solvent as well as solubilizing long peptide chains more effectively. Much more effort will have to be expended before the potential of less hydrophobic supports can be evaluated. Enzymic cleavage from a hydrophilic support is theoretically feasible, but the close approach of enzyme and polymeric substrate may be difficult to accomplish.

In summary, the effects of the physical properties of the polymeric support on the chemical reactions which are transpiring cannot be underestimated and, certainly, no longer ignored. Because the physical properties are being dynamically affected by the elongating chains, sequence-dependent difficulties may arise. The heterogeneous sites present on the polymeric support may well change in accessibility during a synthetic procedure. Other difficulties may arise from distribution of the solute between the solvent and the solvent-polymer phases.

VII. AUTOMATION

One of the motivating considerations behind the development of solid phase synthesis (Merrifield, 1962) was the potential for automation that such a procedure would have. The basic requirements for an automated system are shown in Fig. 23. The first machine for peptide synthesis was constructed by Merrifield and co-workers (1966). As might have been expected, many variations on the mechanical details have been developed. Modifications by Robinson (1967) have been primarily in the use of individual solenoid valves rather than rotary selector valves and in the introduction of a monitoring method. Loffet and Close (1968) have used an electromechanical programmer and a nitrogen delivery system rather than a metering pump. Brunfeldt and co-workers (1968) developed the prototype for the first commercial model which has a punch-tape control unit with a great deal of flexibility, as well as a delivery system under nitrogen pressure which is photoelectrically controlled. Mansveld et al. (1968) have constructed an apparatus similar to that of Loffet and Close (1968), but which features an x-y matrix plug board to maintain flexibility in the control unit. Other units with varying degrees of modification have been constructed and are presently in use in various laboratories (Sugiyama et al., 1969).

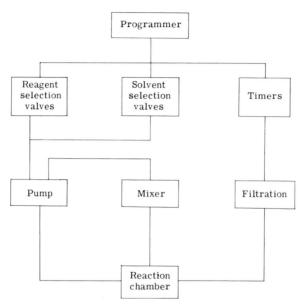

Fig. 23. Basic requirements for an automated apparatus for solid phase synthesis.

VIII. SIMILAR DEVELOPMENTS IN CONVENTIONAL SYNTHESIS

(a) The need for improvements to peptide synthesis in the traditional approach has, generally, been recognized, and a variety of modifications have been suggested. Most of these techniques involve some improvement which facilitates purification or attempts to obviate the need for isolation of intermediates. Naturally, these schemes are open to the same sorts of criticism which have been leveled at the solid phase approach. The work of Goodman and Steuben (1959), Knorre and Shubina (1963, 1965), Medzihradszky and Radoczy (1966), Sheehan et al. (1965), and Laufer and Blout (1967) would fall in this general category and have been recently reviewed (Merrifield, 1969a).

(b) The use of protecting groups which are charged and offer a basis for purification have some overall resemblance to the solid phase technique, although they logically are only extensions of the easy purification scheme and use homogeneous reactions. Young and co-workers (see Camble et al., 1968) used the weakly basic 4-picolyl ester for protection of the C-terminal carboxyl residue. The coupling in solution with a carbobenzoxyamino acid with DCCI in THF was followed by binding to sulfoethyl Sephadex H^+ form and elution with triethylamine after thorough washing. Deprotection by HBr allowed the synthesis to be recycled. The 4-picolyl group was removed by hydrogenation. The general scheme is outlined in Fig. 24, and was applied to the synthesis of angiotensin resulting in a 29% overall yield (Garner and Young, 1969).

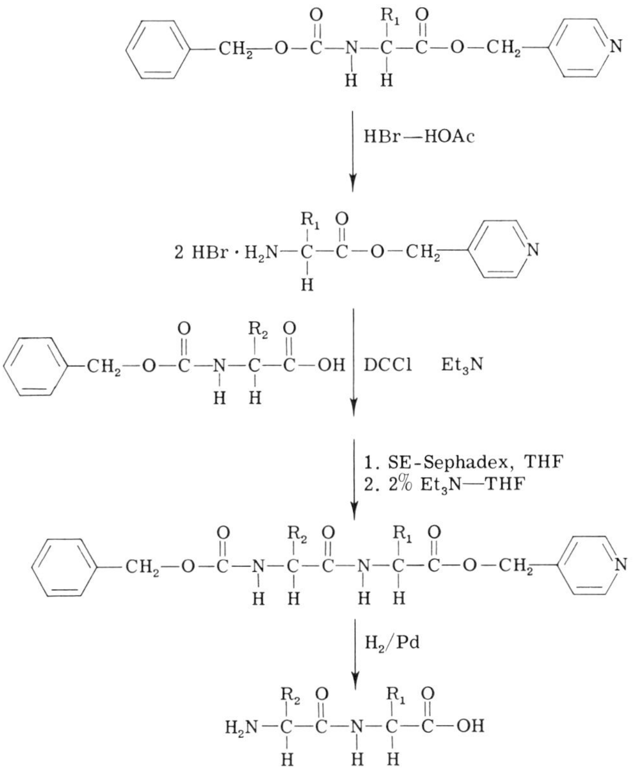

Fig. 24. Use of a charged protecting group as a means of simplifying purification procedures. The use of the 4-picolyl group according to Camble et al. (1968) is illustrated.

Wieland and Racky (1968) have used the ester of p-dimethylaminoazobenzyl alcohol and the C-terminal residue in a similar manner. Such Boc amino acid esters were deprotected with TFA and coupled with Boc-protected amino acids by DCCI. The coupled derivative was purified by binding to sulfoethyl-Sephadex, washing with methanol, and elution with 2% triethylamine in methanol. The colored azo ester was removed by catalytic hydrogenation. H-Pro-Ala-Phe-OH was synthesized in 45% yield to illustrate the application of this novel carboxyl-protection group.

IX. DIFFICULTIES IN ANALYSIS

In addition to the several benefits in the use of solid supports for peptide synthesis that have already been discussed, several disadvantages are also recognized. Quantitation is of primary importance to a method that depends on attaining essentially complete reaction at each step and rather severe limitations on analytical techniques are imposed by the physical state of the polymer-bound peptide. It is essential to develop an accurate, routine method for determining the extent of the deprotection and coupling reactions. If these two reactions could be analyzed quickly and precisely the full advantages of the automated procedures could be realized.

Amino acid analysis has been the routine method of ascertaining the success of the coupling reactions. Difficulties in complete cleavage of polymer-bound peptide led to the use of hydrolysis mixtures such as HCl/HOAc (Merrifield, 1964a), HCl/dioxane (G. R. Marshall and Merrifield, 1965), and, more recently, HCl/proprionic acid (Scotchler et al., 1970). Amino acid analysis is time consuming, only relatively accurate, and difficult to interpret after more than one residue of a given type has been introduced. For this reason, several other possible methods have been introduced.

Mass spectrometry has been used on cleaved peptides by Bayer et al. (1968) in order to detect error sequences. Weygand and Obermeier (1968) coupled quantitative mass spectrometry with Edman degradation in order to determine the kinetics of acylation of the free amino groups using the phenacyl polymer. This is the most extensive study of the effects of various reaction conditions on the coupling reaction, and the results, although not directly applicable to supports other than the phenacyl resin, are certainly suggestive.

Gut and Rudinger (1968) have developed a reaction vessel whose liquid phase can be continuously monitored by optical measurements. The uptake of highly colored derivatives such as the Nps amino acids onto the polymer can be determined as well as the release of the Nps group during deprotection. Certain other groups of practical value do not lend themselves to this method, unfortunately.

Krumdieck and Baugh (1969) have applied the radioactively labeled Boc groups which were suggested by Deer (1968) as a means of measuring both acylation and deprotection. They were able to use it to full advantage since polyglutamate derivatives only require the use of one such derivative. The method, however, suffers the disadvantage, as do many others which have been proposed, of requiring a sample of the polymeric material to be removed for analysis.

A nondestructive method was proposed by Esko et al. (1968) based on Schiff's base formation between 2-hydroxy-1-naphthaldehyde and the free amine, followed by displacement with another amine, e.g. benzylamine, and determination of the soluble aldimine chromophore. The results obtained by this procedure to determine coupling kinetics demonstrated the kinetic effects of the steric influence of side chains, i.e., rate of Boc glycine > Boc alanine > Boc leucine. The primary disadvantage of this approach, at least as demonstrated, was the delay due to the 12-hour Schiff's base formation.

Assay of amine hydrochloride after deprotection with hydrogen chloride in various solvents has been a useful method. Titration of the chloride in the triethylamine hydrochloride resulting from the neutralization gives an indication of the free amino groups on the polymer. Work by Okuda and Zahn (1969) showed very nicely that addition of a second basic group such as an imidazole group could be readily determined by the extra chloride bound. Probable cleavage of the growing polymer chain from the support, using 1 M HCl in glacial acetic acid for deprotection in the synthesis of apoferrodixin (Bayer et al., 1968), was demonstrated in this manner. Dorman (1969) has developed a procedure for measuring coupling based on binding a chloride ion as a salt through a pyridine hydrochloride wash. The chloride was removed by base and titrated. Dorman suggests that these procedures may prove limited in that many of the more labile protecting groups such as Nps, Bpoc, and Bip may be cleaved to some extent in the presence of pyridine hydrochloride.

Infrared spectroscopy has proved to be particularly useful in ascertaining modification to the polymeric support. This was originally demonstrated by Letsinger and co-workers (1964) in their studies of the popcorn polymer and the preparation of a carbobenzoxy derivative. Recent examples of its use are in the preparation of the t-butyloxycarbonyl hydrazide support by Wang and Merrifield (1969a,b,c) and in the monitoring of activation of the dual function supports by oxidation (Flanigan and Marshall, 1970).

Brunfeldt and co-workers (1969b) described the application of nonaqueous potentiometric titration to the problem of assaying free polymeric amino groups. Using electrodes built into a special reaction vessel, which had provisions for stirring, the free amino groups were titrated with 0.1 M perchloric acid in HOAc. Several questions regarding general applicability, adequate sensitivity, and electrode reliability must be answered before this interesting approach can be evaluated.

Beyerman (1971) reported the use of a 17% solution of bromocresyl purple to indicate free amine on the polymer. Kaiser et al. (1970) have used ninhydrin for similar qualitative tests.

The development of rapid and accurate analysis of the extent of reactions which can be used in conjunction with the automated procedure is essential. Careful synthesis requires monitoring of the reactions, and the difficulties and delays encountered in the more common methods severely hinder the potential of the technique.

X. RESULTS

The interest in solid phase peptide synthesis has been strong as exemplified in Table III by the partial list of products made by this method. The table includes peptides not contained in either the "Handbook of Biochemistry," 1970, or in a recent review (Merrifield, 1969a). In addition, the basic idea of solid phase synthesis is of general application to other classes of polymers. Significant efforts to apply this approach to polynucleotides, polyamides, polysacchardies, and desipeptides have been made and have been reviewed (Merrifield, 1969a).

TABLE III

SOLID PHASE PEPTIDE SYNTHESES

Parent compound	No. of residues	Reference
I. Synthesis of complete molecule		
A. Gastrin tetrapeptide	4	Shchukina et al. (1968); Pietta and Marshall (1970)
B. Bradykinin-potentiating factor	5	Greene et al. (1969)
C. Angiotensin II	8	Marshall and Merrifield (1965); Jorgensen et al. (1969); Khosla et al. (1967a, 1968); Marshall (1970); Ovchinnikov et al. (1969); Park et al. (1967); Semkin et al. (1968)
D. Polyglutamyl folic acid	8	Krumdieck and Baugh (1969)
E. Bradykinin	9	Merrifield (1964b, c); Patchornik et al. (1967); Stewart and Woolley (1965a, 1966); Stewart et al. (1967); Fridkin et al. (1968a)
F. Oxytocin	9	Manning (1968); Takashima et al. (1968, 1969); Baxter et al. (1969); Bayer and Hagenmaier (1968); Sawyer et al. (1969); Bayer (1971); Beyerman et al. (1967, 1968b); Beyerman and Intveld (1969); Beyerman and Maassen van den Brink-Zimmermanova (1968)
G. Vasopressin	9	Meienhofer and Sano (1968)
H. Angiotensin I	10	Thampi et al. (1968)
I. Antamanide	10	Bayer and Koenig (1969); Wieland et al. (1969a)
J. Gramicidin	10	Halstrom and Klostermeyer (1968); Klostermeyer (1968); Ovchinnikov et al. (1969)

TABLE III (Continued)

	Parent compound	No. of residues	Reference
K.	Valinomycin	12	Gisin et al. (1969)
L.	Renin tetradecapeptide substrate	14	Skeggs et al. (1969)
M.	Fibrinopeptide A	16	Johnson and May (1969)
N.	Polistis kinin	18	Stewart (1968)
O.	Insulin		
	1. A-chain	21	Marglin and Merrifield (1966); Hornle (1967); Hornle et al. (1968); Marglin and Cushman (1967); Weber et al. (1968); Zahn et al. (1967a, b)
	2. B-chain	30	Marglin and Merrifield (1966); Zahn et al. (1967a, b); Merrifield (1969a); Weber and Weitzel (1968); Witzel et al. (1968a, b)
P.	Apoferrodoxin	55	Bayer et al. (1968)
Q.	Cytochrome c	104	Robinson (1967); Robinson and Kamen (1968); Nanzyo and Sano (1968); Sano and Kurihara (1969); Sano et al. (1968)
R.	Ribonuclease	124	Gutte and Merrifield (1969)
II. Synthesis of partial sequence			
A.	Eledoisin	11	Halstrom et al. (1969); Shchukina and Skylarov (1966)
B.	Melittin	26	Miura et al. (1969)
C.	Glucagon	29	Marshall (1967)
D.	ACTH	39	Blake and Li (1968)
E.	Parathyroid hormone	83	Wang and Merrifield (1969b)
F.	Lysozyme	129	Jolles and Jolles (1968)
G.	Hemoglobin β-chain	146	Chillemi and Merrifield (1969)
H.	Staphylococcal nuclease	149	Anfinsen et al. (1967); Ohno et al. (1969) Ontjes and Anfinsen (1969a, b)
I.	Myoglobin	153	Givas et al. (1968)
J.	Tobacco mosaic virus coat protein	158	Benjamini et al. (1968a, b; 1969) Stewart et al. (1966) Young et al. (1967; 1958a, b)
K.	Casein	~160	Ney and Polzhofer (1968)
L.	Chymotrypsin	245	Mizoguchi and Woolley (1967); Woolley (1966)
M.	Collagen	~1000	Hutton et al. (1968); Kettman et al. (1967); Schoellmann (1967)

XI. CONCLUSIONS AND SUMMARY

Solid phase synthesis arose out of a need to streamline and simplify the techniques of classical synthesis. At present, the method compliments the classical approach and offers a viable alternative to traditional solution reactions. Solid phase synthesis, in its adolescence, is still experiencing maturation pains, but the future appears bright. The proliferation of special purpose supports, more labile bonds to the polymer, investigation of various physical forms for the support as well as techniques for monitoring the extent of reaction can only improve what is already a very useful technique.

With the development of methods for removal of fully protected peptides from the resin and for their reattachment to the polymeric support, the blend of solid phase synthesis with traditional techniques appears possible. Rapid, accurate methods of monitoring the extent of coupling reactions should enable the facile modification of conditions and help to overcome sequence-dependent problems. Alternatively, removal of the growing chain from the support, when difficulties are indicated by the monitoring technique, would allow characterization and purification of intermediates. Reattachment of the purified peptide to the polymer would allow additional synthetic steps until a further purification step was indicated. Although speculative, these appear to be new directions which solid phase peptide synthesis will follow.

References

Albertsson, P. A. (1960). "Partition of Particles and Macromolecules." Wiley, New York.
Anfinsen, C. B., Ontjes, D., Ohno, M., Corley, L., and Eastlake, A. (1967). *Proc. Nat. Acad. Sci. U.S.* **58**, 1806.
Axén, R., and Porath, J. (1964). *Acta. Chem. Scand.* **18**, 2193.
Axén, R., Porath, J., and Ernback, S. (1967). *Nature (London)* **214**, 1302.
Backer, T. A., and Offord, R. E. (1969). *Biochem. J.* 3P.
Baxter, J. W. M., Manning, M., and Sawyer, W. H. (1969). *Biochemistry*, **8**, 3592.
Bayer, E. (1971). *Peptides, Proc. Eur. Symp., 10th, 1969* (in press).
Bayer, E., and Hagenmeier, H. (1968). *Tetrahedron Lett.*, p. 2037.
Bayer, E., and Koenig, W. A. (1969). *J. Chromatogr. Sci.* **7**, 95.
Bayer, E., Jung, G., and Hagenmeier, H. (1968). *Tetrahedron* **24**, 4833.
Bayer, E., Jung, G., Halåsz, I., and Sebastian, I. (1970b). *Tetrahedron Lett.* No. 51, p. 4503.
Bayer, E., Eckstein, H., Hägele, K., Kœnig, W. A., Brüning, W., Hagenmeier, H., and Parr, W. (1970a). *J. Amer. Chem. Soc.* **92**, 1735.
Belleau, B., and Malek, G. (1968). *J. Amer. Chem. Soc.* **90**, 1651.
Ben-Ishai, D., and Berger, A. (1952). *J. Org. Chem.* **17**, 1564.
Benjamini, E., Shimizu, M., Young, J. D., and Leung, C. Y. (1968a). *Biochemistry* **7**, 1253.

Benjamin, E., Shimizu, M., Young, J. D., and Leung, C. Y. (1968b). *Biochemistry* **7**. 1261.
Benjamini, E., Shimizu, M., Young, J. D., and Leung, C. Y. (1969). *Biochemistry* **8**, 2242.
Beyerman, H. C. (1971). *Peptides, Proc. Eur. Symp., 10th, 1969* (in press).
Beyerman, H. C., and Intveld, R. A. (1969). *Rec. Trav. Chim. Pays-Bas* **88**, 1019.
Beyerman, H. C., and Maassen van den Brink-Zimmermannova, H. (1968). *Rec. Trav. Chim. Pays-Bas* **87**, 1196.
Beyerman, H. C., van Zoest, W. J., and Boers-Boonekamp, C. A. M. (1967). *Peptides Proc. Eur. Symp., 8th, 1966*, p. 117.
Beyerman, H. C., Hindriks, H., and de Leer, E. W. B. (1968a). *Chem. Commun.* p. 1668.
Beyerman, H. C., Boers-Boonekamp, C. A. M., and Massen van den Brink-Zimmermannova, H. M. (1968b). *Rec. Trav. Chim. Pays-Bas* **87**, 257.
Blake, J., and Li, C. H. (1968). *J. Amer. Chem. Soc.* **90**, 5882.
Bodanszky, M., and Bath, R. J. (1969). *Chem. Commun.* p. 1259.
Bodanszky, M., and Ondetti, M. A. (1966). "Peptide Synthesis." Wiley (Interscience) New York.
Bodanszky, M., and Sheehan, J. T. (1964). *Chem. Ind. (London)* p. 1423.
Bodanszky, M., and Sheehan, J. T. (1966). *Chem. Ind. (London)* p. 1597.
Boissonnas, R. A., Guttmann, S., and Jaquenoud, P. A. (1960). *Helv. Chim. Acta* **43**, 1349.
Bondi, E., Fridkin, M., and Patchornik, A. (1968). *Isr. J. Chem.* **6**, 22.
Brunfeldt, K. (1971). *Peptides, Proc. Eur. Symp. 10th, 1969* (in press).
Brunfeldt, K., Halstrom, J., and Roepstorff, P. (1968). *Peptides, Proc. Eur. Symp., 9th, 1968*, p. 194.
Brunfeldt, K., Halstrom, J., and Roepstorff, P. (1969a). *Acta Chem. Scand.* **23**, 2830.
Brunfeldt, K., Roepstorff, P., and Thomsen, J. (1962b). *Acta Chem. Scand* **23**, 2906.
Camble, R., Garner, R., and Young, G. T. (1968). *Nature (London)* **217**, 247.
Camble, R., Garner, R., and Young, G. T. (1969). *J. Chem. Soc. London* p. 1911.
Chillemi, F., and Merrifield, R. B. (1969). *Biochemistry* **8**, 4344.
Deer, A. (1966). *Angew. Chem.* **78**, 1064.
Deer, A. (1968). Mann Research Laboratories Catalog.
Denkewalter, R. G., and Hirschmann, R. F. (1969). *Amer. Sci.* **57**, 389.
Dijkstra, A., Billiet, H. A., van Doninck, A. H., van Velthuyzen, H., Maat, L., and Beyerman, H. C. (1967). *Rec. Trav. Chim. Pays-Bas* **86**, 65.
Dorman, L. C. (1969). *Tetrahedron Lett.* No. 28, p. 2319.
Dorman, L. C., and Love, J. (1969). *J. Org. Chem.* **34**, 158.
Du Vigneaud, V., Ressler, C., Swan, J. M., Roberts, C. W., Katsoyannis, P. G., and Gordon, S. (1953). *J. Amer. Chem. Soc.* **75**, 4879.
Ebihara, H., and Kishida, Y. (1969). *Nippon Kagaku Zasshi* **90**, 819.
Epand, R. F., and Scheraga, H. A. (1968). *Biopolymers* **6**, 1551.
Esko, K., Karlsson, S., and Porath, J. (1968). *Acta. Chem. Scand.* **22**, 3342.
Felix, A. M., and Merrifield, R. B. (1970). *J. Amer. Chem. Soc.* **92**, 1385.
Ferriere, N. (1966). *Sci. Progr., Nature* p. 127.
Flanigan, E., and Marshall, G. R. (1970). *Tetrahedron Lett.* **27**, 2403.
Fraefel, W., and du Vigneaud, V. (1970). *J. Amer. Chem. Soc.* **92**, 1030.
Fridkin, M., Patchornik, A., and Katchalski, E. (1965a). *Isr. J. Chem.* **3**, 69P.
Fridkin, M., Patchornik, A., and Katchalski, E. (1965b). *J. Amer. Chem. Soc.* **87**, 4646.
Fridkin, M., Patchornik, A., and Katchalski, E. (1966). *J. Amer. Chem. Soc.* **88**, 3164.
Fridkin, M., Patchornik, A., and Katchalski, E. (1968a). *J. Amer. Chem. Soc.* **90**, 2953.
Fridkin, M., Patchornik, A., and Katchalski, E. (1968b). *Polym. Prepr., Amer. Chem. Soc., Div. Polym. Chem.* **9**, 221.

Fridkin, M. (1971). *Peptides, Proc. Eur. Symp., 10th, 1969* (in press).
Garner, R., and Young, G. T. (1969). *Nature (London)* **222**, 178.
Garner, R., Schafer, D. J., Watkins, W. B., and Young, G. T. (1968). *Peptides, Proc. Eur. Symp., 9th, 1968* p. 145.
Garson, L. R. (1968). *Diss. Abstr. B* **28**, 3219.
Gisin, B. F., Merrifield, R. B., and Tosteson, D. C. (1969). *J. Amer. Chem. Soc.* **91**, 2691.
Givas, Sister J., Centeno, E. R., Manning, M., and Sehon, A. H. (1968). *Immunochemistry* **5**, 314.
Goodman, M., and Stueben, K. C. (1959). *J. Amer. Chem. Soc.* **81**, 3980.
Grahl-Nielsen, O., and Tritsch, G. L. (1969). *Biochemistry* **8**, 187.
Green, B., and Garson, L. R. (1969). *J. Chem. Soc., London* p. 401.
Greene, L. J., Stewart, J. M., and Ferreira, S. H. (1969). *Pharm. Res. Commun.* **1**, 2.
Gut, V., and Rudinger, J. (1968). *Peptides, Proc. Eur. Symp., 9th, 1968* p. 185.
Guttman, S., and Boissonas, R. A. (1959). *Helv. Chim. Acta* **42**, 1257.
Gutte, B., and Merrifield, R. B. (1969). *J. Amer. Chem. Soc.* **91**, 501.
Halasz, I., and Sebastian, I. (1969). *Angew. Chem., Int. Ed. Engl.* **8**, 453.
Halpern, B., and Nitecki, D. E. (1967). *Tetrahedron Lett.* p. 3031.
Halpern, B., Chew, L., Close, V. A., and Patton, W. (1968). *Tetrahedron Lett.* No. 49, p. 5163.
Halstrom, J. (1967). *Dan. Kemi* **48**, 59.
Halstrom, J., and Klostermeyer, H. (1968). *Justus Liebigs Ann. Chem.* **715**, 208.
Halstrom, J., Brunfeldt, K., Thomsen, J., and Kovacs, K. (1969). *Acta Chem. Scand.* **23**, 2335.
Hardy, P. M., Rydon, H. N., and Thompson, R. C. (1968). *Tetrahedron Lett.* No. 21, p. 2525.
Harrison, I. T., and Harrison, S. (1967). *J. Amer. Chem. Soc.* **89**, 5723.
Hiskey, R. G., and Adams, J. B., Jr. (1966). *J. Org. Chem.* **31**, 2178.
Hiskey, R. G., and Southard, G. L. (1966). *J. Org. Chem.* **31**, 3582.
Hofmann, K., Lindenmann, A., Magee, M. Z., and Khan, N. H. (1952). *J. Amer. Chem. Soc.* **74**, 470.
Hornle, S. (1967). *Hoppe-Seyler's Z. Physiol. Chem.* **348**, 1355.
Hornle, S., Weber, V. U., and Weitzel, G. (1968). *Hoppe-Seyler's Z. Physiol. Chem.* **349**, 1428.
Horvath, C. G., Preiss, B. A., and Lipsky, S. R. (1967). *Anal. Chem.* **39**, 1422.
Hutton, J. J., Marglin, A., Witkop, B., Kurtz, J., Berger, A., and Udenfriend, S. (1968). *Arch. Biochem. Biophys.* **125**, 799.
Ingwall, R. T., Scheraga, H. A., Lotan, N., Berger, A., and Katchalski, E. (1969). *Biopolymers* **6**, 331.
Inukai, N., Nakano, K., and Murakami, M. (1968). *Bull. Chem. Soc. Jap.* **41**, 182.
Ives, D. A. J. (1968). *Can. J. Chem.* **46**, 2318.
Johnson, B. J. (1969). *J. Org. Chem.* **34**, 1178.
Johnson, B. J., and Jacobs, P. M. (1968). *Chem. Commun.* p. 73.
Johnson, B. J., and May, W. P. (1969). *J. Pharm. Sci.* **58**, 1568.
Johnson, B. J., and Trask, E. G. (1968). *J. Org. Chem.* **33**, 4521.
Jolles, P., and Jolles, J. (1968). *Helv. Chim. Acta* **51**, 980.
Jorgensen, E. C., Windridge, G. C., Patton, W., and Lee, T. C. (1969). *J. Med. Chem.* **12**, 733.
Kaiser, E., Colescott, R. L., Bossinger, C. D., and Cook, P. I. (1970). *Anal. Biochem.* **34**, 595.
Kato, I., and Anfinsen, C. B. (1969). *J. Biol. Chem.* **91**, 6488.

Katsoyannis, P. G., Fukuda, K., Tometsko, A., Suzuki, K., and Tilak, M. A. (1964). *J. Amer. Chem. Soc.* **86**, 930.
Kessler, W., and Iselin, B. (1966). *Helv. Chim. Acta* **49**, 1330.
Kettman, J. R., Benjamini, E., Michaeli, D., and Leung, D. Y. K. (1967). *Biochem. Biophys. Res Commun.* **29**, 623.
Khosla, M. C., Smeby, R. R., and Bumpus, F. M. (1967a). *Biochemistry* **6**, 754.
Khosla, M. C., Smeby, R. R., and Bumpus, F. M. (1967b). *Science* **156**, 253.
Khosla, M. C., Chaturvedi, N. C., Smeby, R. R., and Bumpus, F. M. (1968). *Biochemistry* **7**, 3417.
Kiryushkin, A. A., Ovchinnikov, Y. A., Kozhevnikova, I. V., and Shemyakin, M. M. (1967). *Peptides, Proc. Eur. Symp., 8th, 1966* p. 100.
Klostermeyer, H. (1968). *Chem. Ber.* **101**, 2823.
Klostermeyer, H., Halstrom, J., Kusch, P., Fohles, J., and Lunkenheimer, W. (1967). *Peptides, Proc. Eur. Symp., 8th, 1966* p. 113.
Knorre, D. G., and Shubina, T. N. (1963). *Dokl. Akad. Nauk SSSR*, **150**, 559; *Chem. Abstr.* **59**, 8870 (1963).
Knorre, D. G., and Shubina, T. N. (1965). *Acta Chim. Acad. Sci. Hung.* **44**, 77.
Krumdieck, C. L., and Baugh, C. M. (1969). *Biochemistry* **8**, 1568.
Kung, Y. T., Du, Y. C., Huang, W. T., Chen, C. C., Ke, L. T., Hu, S. C., Jiang, R. O., Chu, S. Q., Niu, C. I., Hsu, J. Z., Chang, W. C., Cheng, L. L., Li, H. S., Wang, Y., Loh, T. P., Chi, A. H., Li, C. H., Shi, P. T., Yie, Y. H., Tang, K. L., and Hsing, C. Y. (1965). *Sci. Sinica* **14**, 1710.
Kunin, R., Meitzner, E., and Bortnick, N. (1962). *J. Amer. Chem. Soc.* **84**, 305.
Kusch, P. (1966). *Kolloid-Z. Z. Polym.* **208**, 138.
Laufer, D. A., and Blout, E. R. (1967). *J. Amer. Chem. Soc.* **89**, 1246.
Laufer, D. A., Chapman, T. M., Marlborough, D. I., Vaidya, V. M., and Blout, E. R. (1968). *J. Amer. Chem. Soc.* **90**, 2696.
Lenard, J., and Robinson, A. B. (1967). *J. Amer. Chem. Soc.* **89**, 181.
Letsinger, R. L., and Jerina, D. M. (1967). *J. Polym. Sci., Part A-1* **5**, 1977.
Letsinger, R. L., and Kornet, M. J. (1963). *J. Amer. Chem. Soc.* **85**, 3045.
Letsinger, R. L., Kornet, M. J., Mahadevan, V., and Jerina, D. M. (1964). *J. Amer. Chem. Soc.* **86**, 5163.
Loffet, A. (1967). *Experientia* **23**, 406.
Loffet, A., and Close, J. (1968). *Peptides, Proc. Eur. Symp., 9th, 1968* p. 189.
Lombardo, M., and Piasio, R. (1968). Personal communication.
Losse, G., and Neubert, K. (1968a). *Z. Chem.* **8**, 228.
Losse, G., and Neubert, K. (1968b). *Z. Chem.* **8**, 387.
Losse, G., Grenzer, W., and Neubert, K. (1968a). *Z. Chem.* **8**, 21.
Losse, G., Madlung, C., and Lorenz, P. (1968b). *Chem. Ber.* **101**, 1257.
Lutz, F., Rembges, H., and Frimmer, M. *Pharmacology* **2**, 271.
Manecke, G., and Haake, E. (1968). *Naturwissenschaften* **55**, 343.
Manning, J. M., and Moore, S. (1968). *J. Biol. Chem.* **21**, 5591.
Manning, M. (1968). *J. Amer. Chem. Soc.* **90**, 1348.
Manning, M., Wuu, T. C., Baxter, J. W. M., and Sawyer, W. H. (1968). *Experientia* **24**, 659.
Mansveld, G. W. H. A., Hindriks, H., and Beyerman, H. C. (1968). *Peptides, Proc. Eur. Symp., 9th, 1968* p. 197.
Marglin, A., and Cushman, S. W. (1967). *Biochem. Biophys. Res. Commun.* **29**, 710.
Marglin, A., and Merrifield, R. B. (1966). *J. Amer. Chem. Soc.* **88**, 5051.
Marshall, D. L., and Liener, I. E. (1969). *158th Nat. Meet., Amer. Chem. Soc.* B219.
Marshall, G. R. (1967). *Advan. Exp. Med. Biol.* **2**, 48.
Marshall, G. R. (1968). *155th Nat. Meet., Amer. Chem. Soc.* A-4.

Marshall, G. R. (1970). In "Peptides: Chemistry and Biochemistry" (B. Weinstein and S. Lande, eds.), p. 151. Marcel Dekker, New York.
Marshall, G. R., and Flanigan, E. (1969). Unpublished data.
Marshall, G. R., and Merrifield, R. B. (1965). *Biochemistry* **4**, 2394.
Marshall, G. R., and Merrifield, R. B. (1970). In "CRC Handbook for Biochemistry" (H. A. Sober, ed.), 2nd ed. Chem. Rubber Publ. Co., Cleveland, Ohio.
Medzihradszky, K., and Radoczy, J. (1966). *Peptides, Proc. Eur. Symp., 6th, 1963* p. 49.
Meienhofer, J., and Sano, Y. (1968). *J. Amer. Chem. Soc.* **90**, 2996.
Meienhofer, J., Schnabel, E., Bremer, H., Brinkoff, O., Zabel, R., Sroka, W., Klostermeyer, H., Brandenberg, D., Okuda, T., and Zahn, H. (1963). *Z. Naturforsch. B* **18**, 1120.
Merrifield, R. B. (1962). *Fed. Proc., Fed. Amer. Soc. Exp. Biol.* **21**, 412.
Merrifield, R. B. (1963). *J. Amer. Chem. Soc.* **85**, 2149.
Merrifield, R. B. (1964a). *J. Amer. Chem. Soc.* **86**, 304.
Merrifield, R. B. (1964b). *Biochemistry* **3**, 1385.
Merrifield, R. B. (1964c). *J. Org. Chem.* **29**, 3100.
Merrifield, R. B. (1965a). *Endeavour* **24**, 3.
Merrifield, R. B. (1965b). *Science* **150**, 178.
Merrifield, R. B. (1966). In "Hypotensive Peptides" (E. G. Erdos, N. Back, and F. Sicuteri, eds.), p. 1. Springer, Berlin.
Merrifield, R. B. (1967). *Recent Progr. Horm. Res.* **23**, 451.
Merrifield, R. B. (1968). *Sci. Amer.* **218**, 56.
Merrifield, R. B. (1968). In "CRC Handbook of Biochemistry" (H. A. Sober, ed.), p. C-83. Chem. Rubber Publ. Co., Cleveland, Ohio.
Merrifield, R. B. (1969a). *Advan. Enzymol.* **32**, 221.
Merrifield, R. B. (1969b). *J. Amer. Med. Ass.* **210**, 1247.
Merrifield, R. B., and Corigliano, M. A. (1968). *Biochem. Prep.* **12**, 98.
Merrifield, R. B., and Littau, V. (1968). *Peptides, Proc. Eur. Symp., 9th, 1968* p. 179.
Merrifield, R. B., and Marglin, A. (1967). *Peptides, Proc. Eur. Symp., 8th, 1966* p. 85.
Merrifield, R. B., and Stewart, J. M. (1965). *Nature (London)* **207**, 522.
Merrifield, R. B., Stewart, J. M., and Jernberg, N. (1966). *Anal. Chem.* **38**, 1905.
Millar, J. R., Smith, D. G., Marr, W. E., and Kressman, T. R. E. (1963). *J. Chem. Soc., London* p. 218.
Miller, A. W., and Stirling, C. J. M. (1968). *J. Chem. Soc., London* p. 2612.
Mitchell, A. R., Gupta, S. K., and Roeske, R. W. (1969). *Abstr. 158th Nat. Meet., Amer. Chem. Soc.* OR 21.
Miura, Y., Toyama, M., and Seto, S. (1969). *Sci. Rep. Tohoku Univ.* **20**, 41.
Mizoguchi, T., and Woolley, D. W. (1967). *J. Med. Chem.* **10**, 251.
Mizoguchi, T., Levin, G., Woolley, D. W., and Stewart, J. M. (1968). *J. Org. Chem.* **33**, 903.
Mizoguchi, T., Shigezane, K., and Takamura, N. (1969). *Chem. Pharm. Bull.* **17**, 411.
Najjar, V. A., and Merrifield, R. B. (1966). *Biochemistry* **5**, 3765.
Nanzyo, N., and Sano, S. (1968). *J. Biol. Chem.* **243**, 3431.
Nefkins, G. H. L. (1962). *Nature (London)* **193**, 974.
Ney, K. H., and Polzhofer, K. P. (1968). *Tetrahedron* **24**, 6619.
Ohno, M., and Anfinsen, C. B. (1967). *J. Amer. Chem. Soc.* **89**, 5994.
Ohno, M., Eastlake, A., Ontjes, D. A., and Anfinsen, C. B. (1969). *J. Amer. Chem. Soc.* **91**, 6842.
Okuda, T. (1968). *Naturwissenschaften* **14**, 209.
Okuda, T., and Zahn, H. (1969). *Makromol. Chem.* **121**, 87.

Omenn, G. S., and Anfinsen, C. B. (1968). *J. Amer. Chem. Soc.* **90**, 6572.
Ondetti, M. A., Deer, A., Sheehan, J. T., Pluscec, J., and Kocy, O. (1968). *Biochemistry* **7**, 4069.
Ontjes, D. A., and Anfinsen, C. B. (1969a). *Proc. Nat. Acad. Sci. U.S.* **64**, 428.
Ontjes, D. A., and Anfinsen, C. B. (1969b). *J. Biol. Chem.* **244**, 6316.
Ovchinnikov, Y. A., Kiryushkin, A. A., and Kozhevnikova, I. V. (1969). *J. Gen. Chem. USSR* **38**, 2546.
Park, W. K., Smeby, R. R., and Bumpus, F. M. (1967). *Biochemistry* **6**, 3458.
Patchornik, A. (1968). *Advan. Exp. Med. Biol.* **2**, 11–17.
Patchornik, A., Fridkin, M., and Katchalski, E. (1967). *Peptides, Proc. Eur. Symp., 8th, 1966* p. 91.
Pereira, W., Close, V. A., Jellum, E., Patton, W., and Halpern, B. (1969). *Aust. J. Chem.* **22**, 1337.
Pietta, P. G., and Marshall, G. R. (1970). *J. Chem. Soc. D, London*, p. 650.
Polzhofer, K. P. (1969). *Tetrahedron* **25**, 4127.
Polzhofer, K. P., and Ney, K. H. (1969). *J. Chromatogr.* **43**, 404.
Porath, J., Axén, R., and Ernbäck, S. (1967). *Nature (London)* **215**, 1491
Pourchot, L. M., and Johnson, J. J. (1969). *Org. Prep. Procedures* **1**, 121.
Richards, F. F., Sloane, R. W., Jr., and Haber, E. (1967). *Biochemistry* **6**, 476.
Robinson, A. B. (1967). Thesis, University of California, San Diego, California.
Robinson, A. B., and Kamen, M. D. (1968). *In* "Structure and Function in Cytochromes" (K. Okunuki and M. D. Kamen, eds.), p. 383. University Park Press, Baltimore, Maryland.
Rothe, M., and Dunkel, W. (1967). *J. Polym. Sci., Part B* **5**, 589.
Rothe, M., and Schneider, H. (1966). *Angew. Chem.* **78**, 390.
Rothe, M., Schneider, H., and Dunkel, W. (1966). *Makromol. Chem.* **96**, 290.
Rothe, M., Theysohn, R., Steffen, K. D., Schneider, H., Zamani, M., and Kostrzewa, M. (1969). *Angew. Chem., Int. Ed. Engl.* **8**, 919.
Rudinger, J., and Gut, V. (1967). *Proc., Eur. Symp., 8th, 1966* p. 89.
Ryle, A. P., Leclerc, J., and Falla, F. (1969). *Biochem. J.* 4P.
Sakakibara, S., Shin, M., Fujino, M., Shimonishi, Y., Inoue, S., and Inukai, N. (1965). *Bull. Chem. Soc. Jap.* **38**, 1522.
Sakakibara, S., Shimonishi, Y., Okada, M., and Kishida, Y. (1967a). *Peptides, Proc. Eur. Symp., 8th, 1966* p. 44.
Sakakibara, S., Shimonishi, Y., Okada, M., and Sugihara, H. (1967b). *Bull. Chem. Soc. Jap.* **40**, 2164.
Sakakibara, S., Kishida, Y., Nishizaw, R., and Shimonishi, Y. (1968a). *Bull. Chem. Soc. Jap.* **41**, 438.
Sakakibara, S., Kishida, Y., Kikuchi, Y., Sakai, R., and Kakiuchi, K. (1968b). *Bull. Chem. Soc. Jap.* **41**, 1273.
Sakakibara, S., Nakamizo, N., Kishida, Y., and Yoshimuri, S. (1968c). *Bull. Chem. Soc. Jap.* **41**, 1477.
Sanger, F. (1946). *Biochem. J.* **39**, 507.
Sano, S., and Kurihara, M. (1969). *Happe-Seyler's Z. Physiol. Chem.* **350**, 1183.
Sano, S., Kurihara, M., Nishimura, O., and Yajima, H. (1968). *In* "Structure and Function in Cytochromes" (K. Okunuki and M. D. Kamen, eds.), p. 230. University Park Press, Baltimore, Maryland.
Sawyer, W. H., Wuu, T. C., Baxter, J. W. M., and Manning, M. (1969). *Endocrinology* **85**, 385.
Schlossman, S. F., Herman, J., and Yaron, A. (1969). *J. Exp. Med.* **130**, 1031.

Schoellmann, G. (1967). *Hoppe-Seyler's Z. Physiol. Chem.* **348**, 1629.
Schreiber, J. (1967). *Peptides, Proc. Eur. Symp., 8th, 1966* p. 107.
Schröder, E., and Lübke, K. (1965). "The Peptides," Vol. I, p. 317. Academic Press, New York.
Schwyzer, R., Surbeck-Wegmann, E., and Dietrich, H. (1960). *Chimia* **14**, 366.
Scotchler, J., Lozier, R., and Robinson, A. B. (1970), *J. Org. Chem.* **35**, 3151.
Semkin, E. P., Gafurova, N. D., and Shchukina, L. A. (1967a). *Khim. Prir. Soedin.* **3**, 220.
Semkin, E. P., Smirnova, A. P., and Shchukina, L. A. (1967b). *Zh. Obshch. Khim.* **37**, 1169.
Semkin, E. P., Smirnova, A. P., and Shchukina, L. A. (1968). *J. Gen. Chem. USSR* **38**, 2284.
Shapiro, J. T., Leng, M., and Felsenfeld, G. *Biochemistry* **8**, 3219.
Shchukina, L. A., and Sklyarov, L. Yu. (1966). *Khim. Prir. Soedin.* **2**, 200.
Shchukina, L. A., Semkin, E. P., and Smirnova, A. P. (1967). *Khim. Prir. Soedin.* **3**, 358.
Shchukina, L. A., Sklyarov, L. Yu., and Gorbunov, V. I. (1968). *Khim. Prir. Soedin.* **4**, 120.
Sheehan, J. C., Preston, J., and Cruickshank, J. (1965). *J. Amer. Chem. Soc.* **87**, 2492.
Shemyakin, M. M., Ovchinnikov, Yu. A., Kiryushkin, A. A., and Kozhevnikova, I. V. (1965). *Tetrahedron Lett.* No. 27, p. 2323.
Sieber, P., and Iselin, B. (1968). *Helv. Chim. Acta* **51**, 622.
Skeggs, L. T., Lentz, K. E., Kahn, J. R., Dorer, F. E., and Levine, M. (1969). *Circ. Res.* **25**, 451.
Sklyarov, L. Yu., Gorbunov, V. I., and Shchukina, L. A. (1966). *Zh. Obshch. Khim.* **36**, 2220.
Sklyarov, L. Yu., and Shaskova, I. V., (1969). *Zh. Obshch. Khim.* **39**, 2778.
Smith, R. L., and Dunn, F. W. (1966). *Fed. Proc., Fed. Amer. Soc. Exp. Biol.* **25**, 881.
Southard, G. L., Brooke, G. S., and Pettee, J. M. (1969). *Tetrahedron Lett.* p. 3505.
Stelakatas, G. C., Paganou, A., and Zervas, L. (1966). *J. Chem. Soc., London*, p. 1191.
Stewart, F. H. C. (1967). *Aust. J. Chem.* **20**, 2243.
Stewart, J. M. (1968). *Fed. Proc., Fed. Amer. Soc. Exp. Biol.* **27**, 534.
Stewart, J. M., and Woolley, D. W. (1965a). *Nature (London)* **206**, 619.
Stewart, J. M., and Woolley, D. W. (1965b). *Nature (London)* **207**, 1160.
Stewart, J. M., and Woolley, D. W. (1966). "Hypotensive Peptides" (E. G. Erdos, N. Back, and F. Sicuteri, eds.), p. 23. Springer, Berlin.
Stewart, J. M., and Young, J. D. (1969). "Solid Phase Peptide Synthesis." Freeman, San Francisco, California.
Stewart, J. M., Young, J. D., Benjamini, E., Shimizu, M., and Leung, C. Y. (1966). *Biochemistry* **5**, 3396.
Stewart, J. M., Mizoguchi, T., and Woolley, D. W. (1967). *Abstr., 153rd Meet., Amer. Chem. Soc.* 0-206.
Stryer, L., and Haugland, R. P. (1967). *Proc. Nat. Acad. Sci. U.S.* **58**, 719.
Sugiyama, H., Miura, Y., and Seto, S. (1969). *Yuki Gosei Kagaku Kyokai Shi* **26**, 1010.
Takashima, H., Du Vigneaud, V., and Merrifield, R. B. (1968). *J. Amer. Chem. Soc.* **90**, 1323.
Takashima, H., Fraefel, W., and Du Vigneaud, V. (1969). *J. Amer. Chem. Soc.* **91**, 6182.
Takashima, H., Hruby, V. J., and Du Vigneaud, V. (1970). *J. Amer. Chem. Soc.* **92**, 677.
Tesser, G. I., and Ellenbroek, B. W. J. (1967). *Peptides, Proc. Eur. Symp., 8th, 1966* p. 124.
Thampi, N. S., Schoellmann, G., Hurst, M. W., and Huggins, C. G. (1968). *Life Sci.* **7**, Part II, 641.
Tilak, M. A. (1968). *Tetrahedron Lett.* No. 60, p. 6323.

Tilak, M. A., and Hollinden, C. S. (1968). *Tetrahedron Lett.* p. 1297.
Tregear, G. W., Catt, K., and Niall, H. D. (1967). *Proc. Roy. Aust. Chem. Inst.* p. 345.
Tripp, S. L. (1968). *Diss. Abstr. B* **28**, 3661.
Tritsch, G. L., and Grahl-Nielsen, O. (1969). *Biochemistry* **8**, 1816.
Vesa, V. (1968). *Usp. Khim.* **37**, 246.
Vesa, V., and Dirvianskyte, N. (1968). *Liet. TSR Mokslu Akad. Darb., Ser. C* **2**, 59.
Visser, S., Roeloffs, J., Kerling, K. E. T., and Havinga, E. (1968). *Rec. Trav. Chim. Pays-Bas* **87**, 559.
Vlasov, G. P., and Bilibin, A. Y. (1969). *Izv. Akad. Nauk SSSR, Ser. Khim.* p. 1400.
Wang, S. S., and Merrifield, R. B. (1969a). 158*th Nat. Meet., Amer. Chm. Soc.* B216.
Wang, S. S., and Merrifield, R. B. (1969b). *Int. J. Protein Res.* **1**, 235.
Wang, S. S., and Merrifield, R. B (1969c). *J. Amer. Chem. Soc.* **91**, 6488.
Weber, V. U., and Weitzel, G. (1968). *Hoppe Seyler's Z. Physiol. Chem.* **349**, 1431.
Weber, V. U., Hornle, S., Grieser, G., Herzog, K. H., and Weitzel, G. (1967). *Hoppe-Seyler's Z. Physiol. Chem.* **348**, 1715.
Weber, V. U., Hornle, S., Kohler, P., Nagelschneider, G., Eisele, K., and Weitzel, G. (1968). *Hoppe-Seyler's Z. Physiol. Chem.* **349**, 512.
Weitzel, G., Weber, U., Hornle, S., and Schneider, F. (1968a). *Peptides, Proc. Eur. Symp., 9th, 1968* p. 171.
Weitzel, G., Weber, U., Hornle, S., and Schneider, F. (1968b). *Peptides, Proc. Eur. Symp., 9th, 1968* p. 222.
Weygand, F. (1968). *Peptides, Proc. Eur. Symp., 9th, 1968* p. 183.
Weygand, F., and Obermeier, R. (1968). *Z. Naturforsch. B* **23**, 1390.
Weygand, F., and Ragnarsson, U. (1966). *Z. Naturforsch., B* **21**, 1141.
Wieland, T., and Birr, C. (1966). *Angew. Chem.* **78**, 303.
Wieland, T., and Birr, C. (1967a). *Peptides, Proc. Eur. Symp., 8th, 1966* p. 103.
Wieland, T., and Birr, C. (1967b). *Chimia* **21**, 581.
Wieland, T., and Racky, W. (1968). *Chimia* **22**, 375.
Wieland, T., Birr, C., and Flor, F. (1969a). *Justus Liebigs Ann. Chem.* **727**, 130.
Wieland, T., Birr, C., and Wissenbach, H. (1969b). *Angew. Chem.* **8**, 764.
Wildi, B. S., and Johnson, J. H. (1968). *155th Nat. Meet., Amer. Chem. Soc.* A-8.
Wolman, Y., Kivity, S., and Frankel, M. (1967). *Chem. Commun.* p. 629.
Woodward, R. B., Olafson, R. A., and Mayer, H. (1961). *J. Amer. Chem. Soc.* **83**, 1010.
Woolley, D. W. (1966). *J. Amer. Chem. Soc.* **88**, 2309.
Yaron, A., and Schlossman, S. F. (1968). *Biochemistry* **7**, J2673.
Young, J. D., Benjamini, E., Stewart, J. M., and Leung, C. Y. (1967). *Biochemistry* **6**, 1455.
Young, J. D., Leung, C. Y., and Rombauts, W. A. (1968a). *Biochemistry* **7**, 2475.
Young, J. D., Benajmini, E., and Leung, C. Y. (1968b). *Biochemistry* **7**, 3113.
Zahn, H., Okuda, T., and Shimonishi, Y. (1967a). *Peptides, Proc. Eur. Symp., 8th, 1966* p. 108.
Zahn, H., Okuda, T., and Shimonishi, Y. (1967b). *Angew. Chem.* **79**, 424.

CHAPTER 4   Sequential Degradation of Peptides Using Solid Supports

GEORGE R. STARK

I. Introduction 171
II. Degradation of Peptides Attached to Solid Supports . 173
III. Degradation of Peptides with an Insoluble Reagent . 179
IV. Properties of Resins Important for Peptide Degradations 184
References 187

I. INTRODUCTION

The amino acid sequence of every protein we know of has been obtained with at least partial use of stepwise degradation of peptides from their N-termini with isothiocyanates. This procedure was first reported by Edman (1950). Although phenylisothiocyanate has been widely used previously, methylisothiocyanate, which is more volatile and much more soluble in aqueous solvents, is becoming increasingly popular. The degradation of a peptide with an isothiocyanate is carried out in 3 major stages, as illustrated for phenylisothiocyanate in Fig. 1. The stages are (I) a coupling reaction, in which the α-amino group of the peptide reacts with the isothiocyanate; (II) a cleavage reaction (after separation of the derivatized peptide from solvents and excess reagent) in which a strong acid, usually anhydrous trifluoroacetic, catalyzes removal of the terminal residue as a thiazolinone, without disruption of internal peptide bonds; and finally, (III) isomerization of the thiazolinone to a more stable thiohydantoin for purposes of identification. The number of times this cycle of reactions can be repeated with a single peptide depends upon the average yield at each cycle.

Several different groups of workers have used the principle of immobilizing the peptide (or protein) as a means of speeding up and simplifying the Edman degradation. Recently, Edman and Begg (1967) have described the "sequenator," an automatic instrument used for degrading proteins. A substance undergoing degradation is held against the wall of a spinning cup while the reagents flow over the sample in turn. Several commerical instruments employing this principle are now available (Sauer et al., 1970; Hermodson et al., 1970). In another method, described by Schroeder (1967), the peptide is immobilized onto a strip of paper

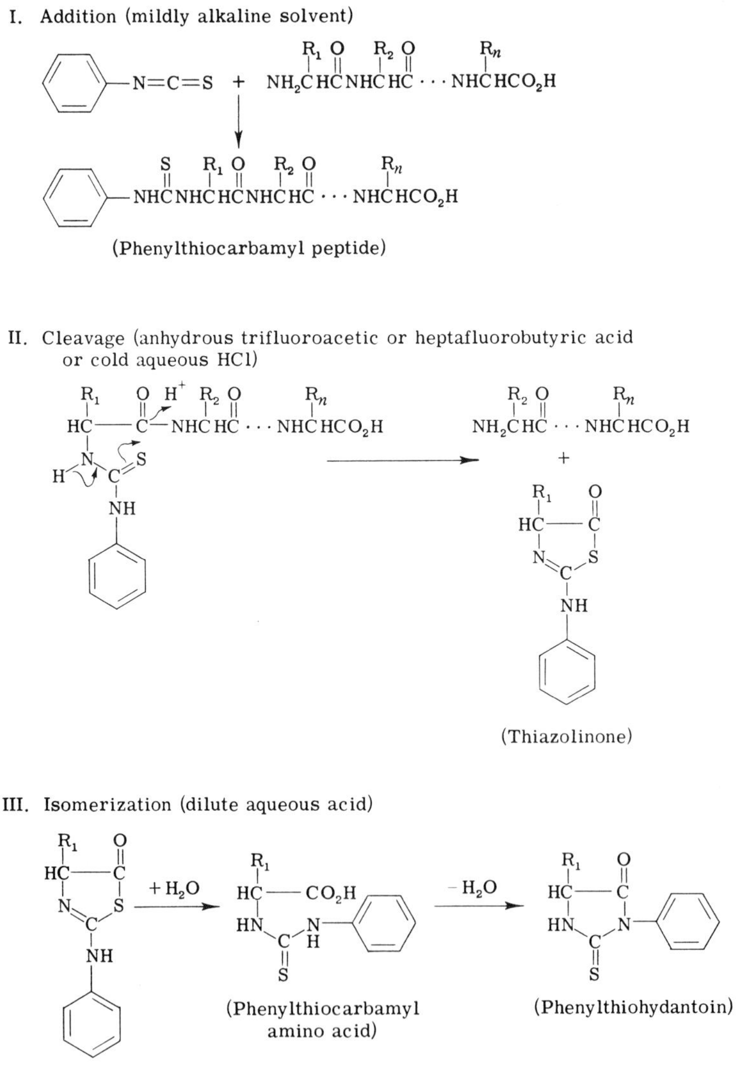

I. Addition (mildly alkaline solvent)

$$\text{Ph}-N=C=S + NH_2\overset{R_1}{\underset{|}{C}}H\overset{O}{\underset{||}{C}}NH\overset{R_2}{\underset{|}{C}}H\overset{O}{\underset{||}{C}}\cdots NH\overset{R_n}{\underset{|}{C}}HCO_2H$$

$$\downarrow$$

$$\text{Ph}-NH\overset{S}{\underset{||}{C}}NH\overset{R_1}{\underset{|}{C}}H\overset{O}{\underset{||}{C}}NH\overset{R_2}{\underset{|}{C}}H\overset{O}{\underset{||}{C}}\cdots NH\overset{R_n}{\underset{|}{C}}HCO_2H$$

(Phenylthiocarbamyl peptide)

II. Cleavage (anhydrous trifluoroacetic or heptafluorobutyric acid or cold aqueous HCl)

(Thiazolinone)

III. Isomerization (dilute aqueous acid)

(Phenylthiocarbamyl amino acid)   (Phenylthiohydantoin)

Fig. 1. Sequential degradation of a peptide with phenylisothiocyanate.

by adsorption. In yet another approach, Haber, Waterfield, and Smith (1968; see Stark, 1970) explored a procedure in which the peptide is deposited as a thin film on glass beads and is then exposed to gaseous reagents. In each of these three cases, extractions and separations, which can be troublesome and time consuming, and which usually result in progressive loss of the peptide, are avoided. Also, it is much easier to enclose the entire system and exclude oxygen when the sample undergoing degradation is immobilized. As Edman et al. have pointed out, it is important to control environmental factors such as oxygen (Ilse and Edman, 1963) and aldehyde contaminants in the solvents (Edman and Begg, 1967). More recently, Hermodson et al. (1970) have found that oxidative breakdown of thiazolinones and thiohydantions is largely prevented by dithioerythritol. Of course, automation also allows degradation to be carried out on a continuous 24-hour schedule. As they do not employ solid supports, these methods are outside the major emphasis of this book and will not be discussed further. They have been treated at somewhat greater length in a recent review (Stark, 1970).

Solid supports have been utilized for sequential degradation of peptides in two distinctly different ways: by immobilizing either the peptide or the isothiocyanate. In the methods of Laursen (Laursen, 1966, Laursen and Bonner, 1970; see also the reports of Schellenberger et al., 1967 and of Dijkstra et al., 1967), Dintzis (1970), and Mross and Doolittle (1970), the C-terminus of the peptide is attached to a resin through a stable amide bond and degradations are then performed with phenyl- or methylisothiocyanate, followed by identification of the thiohydantoin released after each cycle of reactions. In the method of Dowling and Stark (1969), polystyrene, modified with hydrophilic groups, is used as an insoluble support for the reagent. The resin reacts with the α-amino group of a peptide and the peptide is cleaved off subsequently with loss of the N-terminal residue. In this case, the degradation can be followed after each cycle of reactions by determining either the amino acid composition or the new end group of the degraded peptide.

It should be mentioned that the same advantages which make these insoluble supports easy to use make them easy to synthesize: isolation requires only a filtration, and excess reagents are simply washed away.

II. DEGRADATION OF PEPTIDES ATTACHED TO SOLID SUPPORTS

In Laursen's procedure (Laursen and Bonner, 1970), which is summarized in Fig. 2, less than 0.5 μmole of peptide is first treated with t-butyloxycarbonyl azide and N-methylmorpholine in dimethylformamide to block the amino groups (Levy and Carpenter, 1967). Excess reagents and solvents are then removed under high vacuum. Then, the t-butyloxycarbonyl peptide is stirred with a mixture of dimethylformamide and carbonyldiimidazole and added to about 100 mg of a new amino resin. This resin has the structure:

—NH$_2$ (~100%)

CH$_2$NHCH$_2$CH$_2$NH$_2$ (~15%)

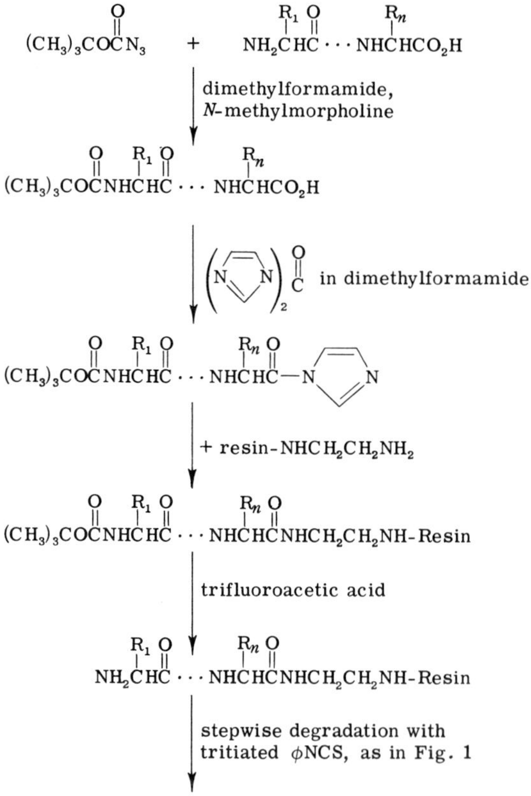

$$(CH_3)_3COCN_3 \quad + \quad NH_2\overset{R_1}{\underset{|}{C}}H\overset{O}{\underset{\|}{C}} \cdots NH\overset{R_n}{\underset{|}{C}}HCO_2H$$

| dimethylformamide,
| N-methylmorpholine

$$(CH_3)_3CO\overset{O}{\underset{\|}{C}}NH\overset{R_1}{\underset{|}{C}}H\overset{O}{\underset{\|}{C}} \cdots NH\overset{R_n}{\underset{|}{C}}HCO_2H$$

| $\left(\underset{N}{\overbrace{}}N\right)_2 \overset{O}{\underset{\|}{C}}$ in dimethylformamide

$$(CH_3)_3CO\overset{O}{\underset{\|}{C}}NH\overset{R_1}{\underset{|}{C}}H\overset{O}{\underset{\|}{C}} \cdots NH\overset{R_n}{\underset{|}{C}}H\overset{O}{\underset{\|}{C}}-N\overbrace{}N$$

| + resin-$NHCH_2CH_2NH_2$

$$(CH_3)_3CO\overset{O}{\underset{\|}{C}}NH\overset{R_1}{\underset{|}{C}}H\overset{O}{\underset{\|}{C}} \cdots NH\overset{R_n}{\underset{|}{C}}H\overset{O}{\underset{\|}{C}}NHCH_2CH_2NH\text{-Resin}$$

| trifluoroacetic acid

$$NH_2\overset{R_1}{\underset{|}{C}}H\overset{O}{\underset{\|}{C}} \cdots NH\overset{R_n}{\underset{|}{C}}H\overset{O}{\underset{\|}{C}}NHCH_2CH_2NH\text{-Resin}$$

| stepwise degradation with
| tritiated ϕNCS, as in Fig. 1

Fig. 2. Attachment of peptide to resin for solid-state degradation according to Laursen's method. (Laursen, 1969, 1970; Laursen and Bonner, 1970).

and is prepared by nitration of chloromethyl polystyrene (2% cross-linked, smaller than 400 mesh), followed by reaction of the chloromethyl groups with ethylenediamine and reduction of the nitro groups with $SnCl_2$. The nitration and reduction are performed according to Chen (1955) [see also Dowling and Stark (1969), Section III of this article, and Chapter 5 by Patterson (this volume)]. The extra aromatic amino groups which remain on the resin after the peptide has been coupled facilitate swelling of the resin in trifluoroacetic acid (Section IV). After the peptide has reacted, the resin is washed with dimethylformamide and methanol. Blocking of the N-terminus of the peptide and coupling of the peptide to the resin are carried out in a reaction cell very similar to the one used for peptide synthesis (Merrifield et al., 1966). The resin is then mixed with about 1 g of glass beads, which prevent channeling due to swelling and shrinking, and the mixture is packed into a reaction column 3 mm i. d. × 100 mm long, thermostated at 40°. The blocking group is removed from the N-terminus of the peptide with trifluoroacetic acid, and the resin-bound peptide is then ready for sequential degradation. Addition of phenylisothiocyanate to the N-terminus of the resin-bound peptide is carried out in N-methylmorpholine buffer, pH 8.0, for 20 minutes. The entire operation has been automated, as illustrated in Figs. 3 and 4. Each cycle takes 2 hours, and all operations are controlled by a programmer.

In order to facilitate identification and quantitiation of the thiohydantoins, Laursen uses tritiated phenylisothiocyanate. The radioactive thiohydantoins released from the peptide after each cycle of degradation are analyzed by thin layer chromatography after dilution of the isotope with a mixture of unlabeled thiohydantoins (Laursen, 1969). Other methods of analysis, such as the one utilizing gas-liquid partition chromatography of phenylthiohydantoins (Pisano and Bronzert, 1969), would also be suitable.

Laursen has observed little or no effect of reagent purity on the degradation, in agreement with the results of Dowling and Stark (1969) (see also Section III). In fact, he simply uses reagent grade methanol, dichloroethane, trifluoroacetic acid, and phenylisothiocyanate without purification. The excess free amino and phenylthiocarbamyl groups on the resin probably absorb aldehydes or oxidants in the solvents.

The results of an 8-cycle degradation on 0.35 μmole of a heptapeptide are shown in Fig. 5. The average yield per cycle is about 97%. Peptides containing as many as 30 residues have been attached to the resin successfully.

Mross and Doolittle (1970) have used a procedure very similar to that of Laursen, but different in some noteworthy details. Their resin is prepared from chloromethylpolystyrene (1% cross-linked, 200–400 mesh) by treatment with ethylenediamine, but without the extra amino groups introduced by nitration of the ring and reduction. Peptides are coupled in aqueous 50% dioxane with the water-soluble reagent 1-ethyl-3-(3-dimethylaminopropyl)carbodiimide. Decapeptides have been coupled to the resin by this method to the extent of 98%, and a peptide of 24 residues has been coupled with efficiencies between 45% and 65%.

Dintzis (1970) uses an approach similar in concept to that of Laursen, but quite different in detail. The support for the peptide is a hydrophilic amino resin, formed from polyacrylamide beads and ethylenediamine by amide exchange (Inman and Dintzis, 1969) (see Fig. 6). The peptide is blocked with 2,6-dinitrobenzene-1,4-disulfonate (an analog of trinitrobenzenesulfonate which gives water-soluble derivatives), then coupled to the resin with a water-soluble diimide. Finally, the blocking group is removed with ammonia. Sequential degradation is carried out in a completely aqueous environment, without the use of any organic solvents. The water-soluble reagent CH_3NCS is coupled to the resin-bound peptide in 5×10^{-3} M phosphate buffer, pH about 8. After excess reagent has been removed with a wash of 1×10^{-3} M HCl, cleavage and

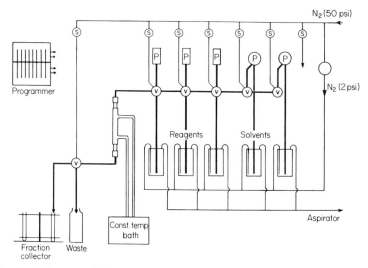

Fig. 3. A diagram of Laursen's apparatus for automated sequential degradation of peptides attached to solid supports. P, pumps; V, valves; S, solenoid valves.

Fig. 4. A photograph of Laursen's apparatus.

conversion to a thiohydantoin are effected simultaneously with cold 6 M HCl. The strong acid has not catalyzed an appreciable amount of nonspecific hydrolysis in any peptides yet tried. High yields have been obtained with some peptides, but with others the degradation has not yet worked well. In the Dintzis procedure, methylisothiocyanate is used, and the course of a degradation is followed by identifying the methylthiohydantoins by thin-layer chromatography (Krivtsov and Stepanov, 1965). Gas-liquid chromatography might also be employed (Waterfield and Haber, 1970).

In all of the methods in which a peptide is to be attached through its α-carboxyl group to an insoluble support, the side chain carboxyls of aspartic and glutamic acid residues present a special problem. Unless they are specifically

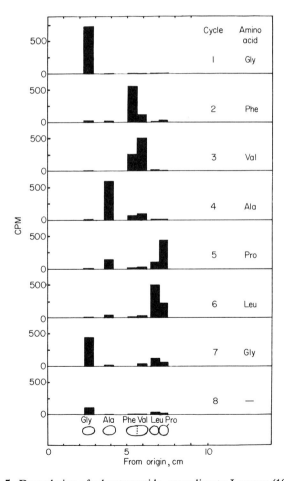

Fig. 5. Degradation of a heptapeptide, according to Laursen (1969).

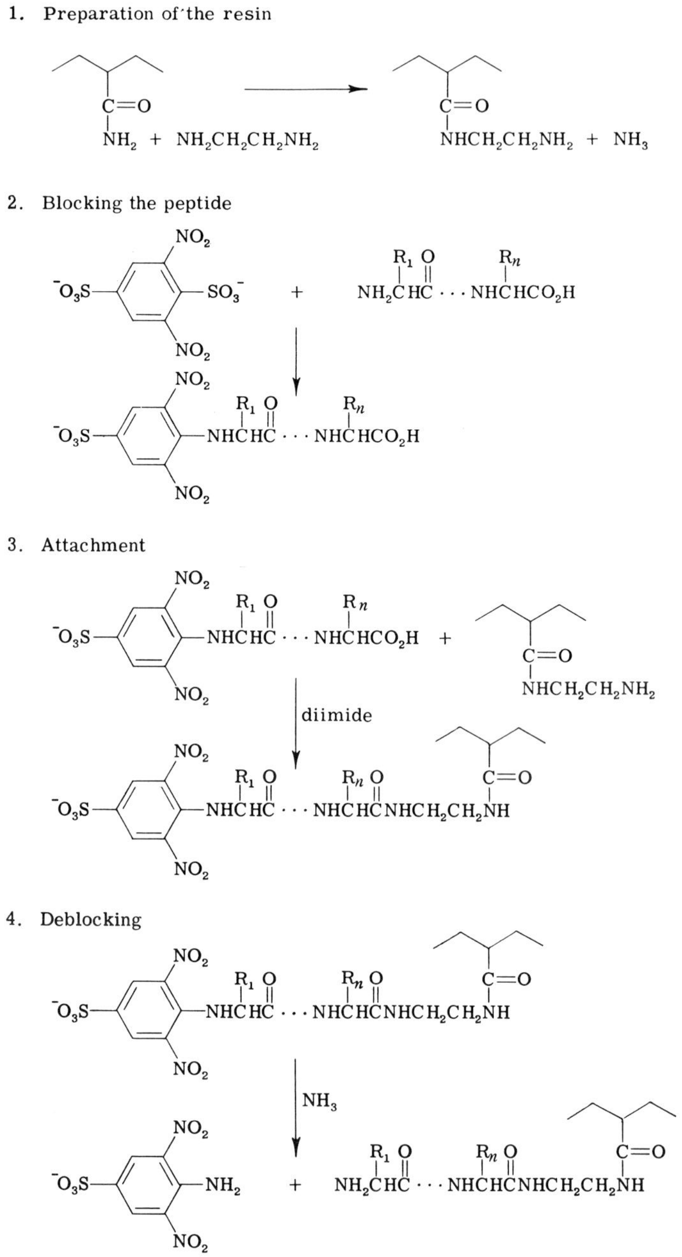

1. Preparation of the resin

2. Blocking the peptide

3. Attachment

4. Deblocking

Fig. 6. Attachment of peptide to resin for solid-state degradation according to Dintzis' method.

blocked, these carboxyls will also become attached to the resin and no thiazolinone or thiohydantoin will be released when the degradation reaches such a residue. If only one aspartic or glutamic acid is present in a peptide, its position might be surmised from such a negative result, although such a procedure seems a bit dangerous. There are two alternative approaches: (a) Block all the free carboxyls on the protein before proteolytic digestion, perhaps with catalysis by a water-soluble carbodiimide. If the blocking group is an alcohol, the α-carboxyl group of the C-terminal peptide might be liberated again by mild alkaline hydrolysis after this peptide has been purified. (b) Obtain, for example, tryptic peptides and block all the carboxyls. Retreatment with trypsin should cleave the blocking group from all the α-carboxyls specifically, except for that of the C-terminal peptide. This latter approach is being actively pursued by Mross and Doolittle (1970), who have coupled all of the carboxyl groups of peptides to glycine amide, using 1-ethyl-3-(3-dimethylaminopropyl)carbodiimide, the same water-soluble carbodiimide they use to couple peptides to resin. Glycine amide was chosen because it reacts quantitatively with the carboxyl groups of peptides [see the results of Hoare and Koshland (1967) with glycine methyl ester], because peptide and amide groups are stable to alkaline conditions (in contrast to esters), and because the glycine amide residue can be removed readily with the same protease used to prepare the peptide. Mross and Doolittle have used fibrinopeptides in their studies, and have cleaved the subterminal arginine–glycine amide bond specifically with trypsin. Since racemization would yield a peptide partially resistant to hydrolysis, the possibility that the amino acid which is to be recognized by the enzyme might be racemized during the diimide-catalyzed coupling is being considered by Mross and Doolittle.

One problem which has been apparent in the use of the sequenator (Edman and Begg, 1967) with small peptides is a rapid drop in yield as a degradation approaches the C-terminus, due to increasing solubility of the peptide in solvents used for extraction of the reagent. One approach to solving this problem has been to alter the sequenator subroutines so as to minimize extraction of the peptide (Sauer et al., 1970). Another approach would be to couple the free α-carboxyl groups of peptides to a high molecular weight polyol or polyamine, using the techniques of Mross and Doolittle (1970).

III. DEGRADATION OF PEPTIDES WITH AN INSOLUBLE REAGENT

The synthesis of an insoluble isothiocyanate (Dowling and Stark, 1969), suitable in its present stage of development for complete degradation of peptides of about 10 residues or less, is illustrated in Fig. 7. Polystyrene beads, cross-linked with 0.25% divinylbenzene and smaller than 150 μ in diameter when dry,

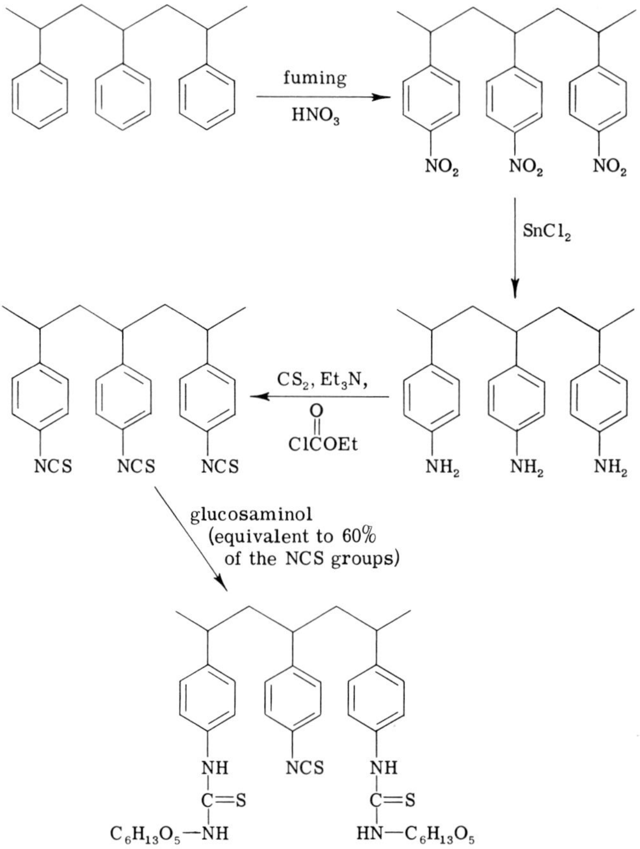

Fig. 7. Synthesis of NCS-polystyrene for peptide degradation.

are first nitrated with fuming nitric acid, then reduced with $SnCl_2 \cdot 2H_2O$ in dimethylformamide according to Chen (1955). Conversion of aminopolystyrene to NCS-polystyrene is accomplished in a two-step reaction with CS_2 and triethylamine, then ethylchloroformate, essentially as described by Garmaise et al. (1958). [A different synthetic route to NCS-polystyrene has been devised by Manecke and Günzel (1968).] The product, NCS-polystyrene, is extremely hydrophobic and must be converted to a more hydrophilic form in order to swell in the aqueous solvents which are required to dissolve peptides. This is accomplished by coupling 60% of the –NCS groups to glucosaminol, thereby introducing a large number of aliphatic hydroxyl groups into the resin. The capacity of the final product is about 1.2 mEq per gram. For further details of its preparation, see Dowling and Stark (1969).

Degradation of a peptide with glucosaminol-NCS-polystyrene is illustrated schematically in Fig. 8. Note that the thiazolinone, which is covalently attached

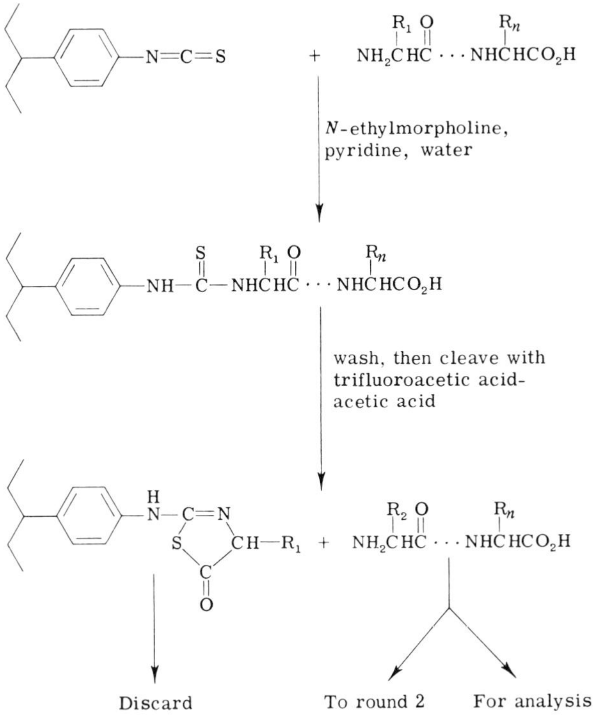

$$\text{Resin-}\text{C}_6\text{H}_4\text{-N=C=S} \;+\; \text{NH}_2\text{CHR}_1\text{CO}\cdots\text{NHCHR}_n\text{CO}_2\text{H}$$

$\downarrow$ N-ethylmorpholine, pyridine, water

$$\text{Resin-}\text{C}_6\text{H}_4\text{-NH-C(=S)-NHCHR}_1\text{CO}\cdots\text{NHCHR}_n\text{CO}_2\text{H}$$

$\downarrow$ wash, then cleave with trifluoroacetic acid–acetic acid

$$\text{Resin-}\text{C}_6\text{H}_4\text{-N(H)-C(=N)-S-C(=O)-CH-R}_1 \;+\; \text{NH}_2\text{CHR}_2\text{CO}\cdots\text{NHCHR}_n\text{CO}_2\text{H}$$

Discard   To round 2   For analysis

Fig. 8. Degradation of a peptide with glucosaminol-NCS-polystyrene.

to the resin, is lost. Degradations can be followed by analyzing a small portion of the degraded peptide either for its amino acid composition after acid- or enzyme-catalyzed hydrolysis, or for its new end group, perhaps by using the sensitive procedure with the fluorescent reagent 1-dimethylaminonaphthalene-5-sulfonyl chloride, as described by Gray (1967). For a typical degradation, a bed of glucosaminol-NCS-polystyrene about 5 cm high is packed into a tube, 0.6 cm in diameter, thermostated at 50°. The column is washed with a solvent composed of pyridine (75 parts), N-ethylmorpholine (5 parts), and water (20 parts). The peptide, dissolved in 200 μl of the same solvent, is placed on the column with gentle nitrogen pressure and allowed to react with the resin for 1 hour. The addition solvent is then washed away with tetrahydrofuran, and cleavage of the peptide is accomplished with a mixture of trifluoroacetic acid (3 parts) and acetic acid (1 part). (The acetic acid is required for good swelling of the resin.) The degraded peptide, recovered by removing the cleavage solvent in a vacuum, is redissolved in addition solvent and, after a small portion of the peptide has been removed for analysis, it is applied to a fresh column of resin for the next cycle of reactions. One cycle takes about 2 hours to perform and several peptides can be degraded in tandem. Automation of the method seems a practical possibility for the future.

The principal advantage of degradations with glucosaminol-NCS-polystyrene, apart from speed and convenience, is that removal of the terminal residue is virtually complete at each stage. This is so because side reactions and incomplete degradation lead to a loss of material rather than to contamination of the product. During the addition reaction, any peptide derivative without a free amino group cannot add to the resin and will be removed in the subsequent washes. Similarly, if the rate of addition is slow at some stage in the degradation, any peptide that has not added to the resin will also be removed. A second purification occurs during the cleavage step, for it is unlikely that any reaction which does not lead to formation of a thiazolinone will allow the product to be released from the resin.

The major current limitation of this procedure is that recoveries of degraded peptide average only about 75% per cycle, and it is this limitation which is primarily responsible for the present restriction of the method to peptides of about 10 residues or less. Even a small improvement in recovery, say to 85% or 90% per stage, would allow the method to be extended to much larger peptides. Several phenomena probably contribute to the poor recoveries. Complete addition may not be obtained in one hour at 50° with all peptides, depending upon the size of the peptide and the nature of the terminal residue. Dowling and Stark (1969) have presented preliminary evidence for a side reaction of unknown nature which blocks the N-terminus in competition with addition to the resin; oxygen and aldehydes, which have been shown to be responsible for side reactions in the Edman degradation with soluble reagents (Ilse and Edman, 1963; Edman and Begg, 1967), do not seem to be responsible in this case. It is probable that recoveries can be improved by further work on the resin itself, to improve the availability of the –NCS groups (see the discussion of the properties of resins in Section IV and the chapter by Patterson elsewhere in this book), and perhaps by a more thorough exploration of optimum conditions for the addition reaction.

Even with recoveries of only about 75% per stage, it has been possible to degrade a dodecapeptide completely with glucosaminol-NCS-polystyrene (Table I). The peptide, derived from bovine catalase, was kindly provided by Dr. Walter Schroeder, who had previously determined its sequence independently by other methods. The data of Table I illustrate several points which have not yet been discussed: (a) if a peptide contains lysine, the ε-amino group must

TABLE I

Degradation of Thr-Thr-Gly-Gly-Asn-Pro-Val-Gly-Asp-Lys-Leu*

Molar ratios of amino acids

Amino acid	Theory	Initial peptide[a]	Round 1[b]	2	3	4	5	6	7	8	9	10	11
Lys	1	0.9	1.0	[c]	[c]	1.0	1.2	[c]	1.0	1.5	[c]	[c]	[c,d]
Asp[e]	2	1.8	2.0	2.3	2.2	2.0	2.0	1.2	1.1	1.7[f]	1.0	0.3	0
Thr	2	1.4	1.0	0.3	0.2	0							
Pro	1	0.9	0.9	1.2	1.3	0.9	0.9	0.8	0	1.7[f]	0.5	0.4	0
Gly	4	3.7	3.5	3.8	3.0	2.1	1.5	1.6	1.2	0.2	0		
Val	1	1.0	0.9	0.9	0.9	1.0	1.1	0.8	1.1	1.0	1.0	1.0	1.0
Leu	1	1.0	1.0	1.0	1.0	1.0	1.0	1.0	1.0	1.0	1.0	33	98
Yield, %			100	18[g]	66	80	65	74	90	86	43	40	[d]
Amount hydrolyzed, %			2.5	2.5	2.5	5.0	5.0	7.5	10	15	20		

* From Dowling and Stark (1969).

[a] This analysis of the initial peptide was made shortly after isolation (November, 1965) by Dr. Walter Schroeder. Our own analysis differed by less than 0.1 residue for every amino acid *except* threonine, which had decreased to 1.1 residues.

[b] This round was carried out with phenylisothiocyanate in solution; the yield at this stage is defined as 100%. Peptide (3.2 μmoles) was put onto the resin for round 2.

[c] Not analyzed; lysine values are somewhat high due to an artifact from the resin.

[d] No hydrolysis.

[e] No attempt was made to distinguish between Asp and Asn.

[f] This particular sample was contaminated by appreciable amounts of serine and alanine, which are not present in the peptide. The high values for aspartic acid and glycine probably reflect the contamination.

[g] The low yield at this point reflects the low value for threonine in the initial peptide; see the text and footnote *a*.

be blocked before degradation with glucosaminol-NCS-polystyrene is begun. This is easily accomplished by carrying out the first cycle of reactions with soluble reagents. In the case illustrated in Table I, the soluble round was followed by 10 rounds on the resin. Note that the C-terminal leucine was analyzed without hydrolysis following round 10. (b) In this particular case, the peptide was not pure, judging from the low recovery of threonine obtained both by Dr. Schroeder in 1965 and by Dowling and Stark in 1969. It appears that the N-terminus of a substantial fraction of the material has been altered. A blocked N-terminus is probably the cause of the exceptionally low recovery of peptide following the first round on the resin. Despite this difficulty, it was possible to achieve complete degradation because the minor fraction of peptide which did have a free N-terminus was separated from the major fraction of impurity during the first degradation with the resin. (c) The recoveries obtained require that an increasing fraction of peptide be used for analysis as a degradation progresses. The procedure used with the dodecapeptide is shown in the last line of Table I.

IV. PROPERTIES OF RESINS IMPORTANT FOR PEPTIDE DEGRADATIONS

In this section I will take the rare opportunity to discuss some of the experiments that failed in attempting to develop resins for use in peptide degradations. Experiments that are unsuccessful lead eventually toward those that are, and some of the lessons learned in attempts at peptide degradation will probably be valuable to those who will seek to employ solid-state methodology for different purposes.

One of the major problems in sequential degradation of peptides with insoluble supports is to maintain the resin in a swollen form in each of the solvents used for addition and cleavage. Swelling is absolutely essential, for the vast majority of functional groups are not accessible to the solvent if the resin collapses. In our earliest experiments, NCS-polystyrene was used. Although it was possible to couple a few small hydrophobic peptides to such a resin in a solvent which contained only a few percent of water, the resin collapsed completely in anhydrous trifluoroacetic acid, and no cleavage could be obtained. Our first successful degradations were accomplished with a resin prepared by treating NH_2-polystyrene with an amount of thiophosgene insufficient to react with all of the amino groups of the resin. The unreacted amino groups became protonated in trifluoroacetic acid and charge-charge repulsion forced the resin to swell during cleavage. However, this resin was uncharged and quite hydrophobic during the coupling reaction in alkaline solvents and could only be used when the addition solvent contained very little water. Unfortunately, most peptides were insoluble in such a solvent. Another problem with this resin is that reaction of resin-bound amino groups with thiophosgene is so rapid relative to diffusion into the resin beads that the distribution of NH_2 and NCS groups is not uniform: most of the NH_2 groups near the exterior of the bead react first.

A generally useful reagent, which remains swollen during both addition and cleavage, was finally prepared from NCS-polystyrene by using some of the NCS groups as points of attachment for the hydrophilic substance glucosaminol (see Fig. 7). The manner in which glucosaminol is added is the most crucial step in the preparation. If glucosaminol hydrochloride in water is added to a stirred suspension of resin in a mixture of pyridine and triethylamine, reaction is complete in less than 15 minutes at room temperature, but the product, although similar in appearance to resin prepared by slow addition of glucosaminol in the presence of N-ethylmorpholine, is very different in its properties. For example, glucosaminol (triethylamine) resin will not react with an appreciable amount of peptide in a solvent containing 20% water, whereas glucosaminol (N-ethylmorpholine) resin reacts with peptides readily under the same conditions. The glucosaminol (triethylamine) resin does react with some peptides when less water is present. The most probable explanation of the difference is that when glucosaminol is present as the free base (as in the presence of excess triethylamine), its rate of reaction with resin-bound NCS groups must be fast relative to its rate of diffusion into the resin beads. The result is a bead which has predominantly $NHC(=S)$-glucosaminol groups near the surface, and predominantly NCS groups near the center. In polar solvents containing 20% water, the peptide cannot penetrate to the hydrophobic interior of the bead, where the NCS groups are to be found, although it probably does penetrate the more hydrophilic exterior. When N-ethylmorpholine replaces triethylamine, the glucosaminol is present predominantly as the hydrochloride, slowing the reaction with NCS groups relative to diffusion and permitting a much more even distribution of NCS groups throughout the bead, so that a greater proportion of them are near the hydrophilic exterior. In the preparation described above, only about 60% of the glucosaminol has reacted with NCS groups after 1.5 hours at room temperature in the presence of N-ethylmorpholine. Triethylamine is added at this point to drive the addition to completion, because prolonged exposure of NCS groups to alkaline conditions in the presence of water results in hydrolysis (Drobnica and Augustin, 1965) and loss of capacity. It is probable that the properties of glucosaminol-NCS-polystyrene could be improved further by more careful control of the mode of addition of glucosaminol. The problems of distribution of functional groups within the resin and relative rates of diffusion and reaction has also been discussed in the context of peptide synthesis by Marshall and Merrifield elsewhere in this book.

Laursen's early experience with resins was in some respects similar to ours. He used aminomethyl-aminoethyl polystyrene, which was acetylated with acetylimidazole after the peptide had been coupled to it. The resulting product was barely penetrable by trifluoroacetic acid; some studies showed that the cyclization step with this material was about 20 times slower than with Edman's sequenator. Laursen (1970) has made the relevant observation that if a 1 : 1 mixture of trifluoroacetic acid and dichloroethane is poured through a dry column of polystyrene, the dichloroethane is selectively adsorbed and pure trifluoroacetic acid comes through. The amino resin now used by Laursen swells up to five times in trifluoroacetic acid, even after many of the amino groups have reacted with phenylisothiocyanate. Presumably, protonation of the ring amino groups and delocalization of the charge into the ring causes repulsion. This modification has allowed him to decrease the temperature of the degradation from 60° to 40°.

In Dintzis' procedure, aqueous conditions are used both during addition and during cleavage and the polyacrylamide support swells well in each step. One factor which may eventually prove troublesome in this method is the use of 6 M HCl for cleavage. Some peptide bonds, notably those involving aspartic acid, are known to be particularly labile in acid (Schultz, 1967) and may cleave partially, expecially during degradation of long peptides. Similarly, the possibility that the resin support might not be completely inert must also be taken into account. Acid hydrolysis of the amide side chains or the methylenebisacrylamide cross-links of polyacrylamide might prove troublesome in some circumstances. It should be noted that reagents other than methylenebisacrylamide can be copolymerized with acrylamide to give a product in which the cross-links are stable to hydrolysis. Such a product may be desirable for some purposes. For example, we have used divinylsulfone successfully in emulsion polymerizations with acrylamide in the presence of tetramethylethylenediamine and persulfate. The resulting bead polymers retain their form after exposure to extreme hydrolytic conditions.

The above comparison of the advantages and disadvantages of polystyrene-based and polyacrylamide-based supports illustrates a typical dilemma: the polystyrene is chemically inert, but ingenuity and striving are required to convert it to a hydrophilic derivative, whereas the acrylamide is already hydrophilic though it may not be inert. Two more dilemmas are (a) How much cross-linking should be used? (too much makes most of the functional groups inaccessible, too little makes the resins sticky and mechanically weak) and (b) How large should the resin beads be? (excessively large size makes the interior functional groups less accessible, while too small a size makes filtration difficult). The way in which solutions to these questions were arrived at is now described for NCS-polystyrene.

Polystyrene was chosen as the supporting matrix for NCS groups because of its resistance to chemical degradation and because it is readily available in stable bead form with a low degree of cross-linking for maximum swelling.

Some experiments were attempted with NCS-Sephadex, prepared according to Axén and Porath (1966), but much of this extremely hydrophilic reagent became soluble in trifluoroacetic acid. Also, the rate of addition of peptide to NCS-Sephadex was much slower than with glucosaminol-NCS-polystyrene and the capacity of the Sephadex was four to five times lower, even after allowing for coupling of 60% of the groups originally present in NCS-polystyrene to glucosaminol. The degree of cross-linking chosen for glucosaminol-NCS-polystyrene, 0.25%, is the lowest degree compatible with physical stability. Several experiments were attempted with 0.1% cross-linked polystyrene, but most of the beads disintegrated during synthesis and the rather mushy final resin was extremely difficult to work with in small columns. Some successful experiments were carried out with 0.25% cross-linked beads initially 25 μ in diameter, but these seemed to have no particular advantages over the larger beads recommended above (minus 100 mesh; i.e., smaller than 150 μ) and were much more difficult to use.

An alternative to a polymeric support that swells in all the required solvents is one that swells in none of them, so that the functional groups are always exposed. Of course, some means must be found to increase surface area drastically. A simple calculation shows that simply decreasing the size of the resin bead is impractical. The ratio of volume to surface is $\frac{4}{3}\pi r^3 / 4\pi r^2 = r/3$. For a very small bead only 16 μ in diameter, the volume to surface ratio is 8/3 μ, or about 27,000 Å. If the functional groups were 10 Å apart, there would be 2700 groups in the interior of such a bead for each one on the surface. Beads as small as 16 μ are extremely difficult to filter. Materials in which the volume to surface ratio has been substantially decreased, such as macroporous polystyrene, "popcorn" polystyrene, and porous glass have been described. Some of these materials have been evaluated in a preliminary way for peptide synthesis (see the article by Marshall and Merrifield elsewhere in this book for a brief description and references) but very little has been done with them in the field of peptide degradation on insoluble supports. Obviously, there are still some new frontiers to be explored.

References

Axén, R., and Porath, J. (1966). *Nature (London)* **210**, 367.
Chen, C. H. (1955). Ph. D. Thesis, Polytechnic Institute of Brooklyn, Brooklyn, New York.
Dijkstra, A., Billiet, H. A., vanDoninck, A. H., van Velthuyzen, H., Maat, L., and Beyerman, H. C. (1967). *Rec. Trav. Chim. Pays-Bas* **86**, 65.
Dintzis, H. M. (1970). Personal communication.
Dowling, L. M., and Stark, G. R. (1969). *Biochemistry* **8**, 4728.
Drobnica, L., and Augustin, J. (1965). *Collect. Czech. Chem. Commun.* **30**, 99.

Edman, P. (1950). *Acta Chem. Scand.* **4**, 277 and 283.
Edman, P., and Begg, G. (1967). *Eur. J. Biochem.* **1**, 80.
Garmaise, D. L., Schwartz, R., and McKay, A. F. (1958). *J. Amer. Chem. Soc.* **80**, 3332.
Gray, W. R. (1967). *Methods Enzymol.* **11**, 469.
Hermodson, M. A., Ericsson, L. H., and Walsh, K. A. (1970). *Fed. Proc., Fed. Amer. Soc. Exp. Biol.* **29**, 728.
Hoare, D. G., and Koshland. D. E., Jr. (1967). *J. Biol. Chem.* **242**, 2447.
Ilse, D., and Edman, P. (1963). *Aust. J. Chem.* **16**, 411.
Inman, J. K., and Dintzis, H. M. (1969). *Biochemistry* **8**, 4074.
Krivtsov, V. F.,and Stepanov, V. M. (1965). *Zh. Obshch. Khim.* **35**, 53, 556, and 982.
Laursen, R. A. (1966). *J. Amer. Chem. Soc.* **88**, 5344.
Laursen, R. A. (1969). *Biochem. Biophys. Res. Commun.* **37**, 663.
Laursen, R. A. (1971). *Eur. J. Biochem.* **20**, 89.
Laursen, R. A., and Bonner, A. G. (1970). *Fed. Proc., Fed. Amer. Soc. Exp. Biol.* **29**, 727.
Levy, D., and Carpenter, F. H. (1967). *Biochemistry* **6**, 3559.
Manecke, G., and Günzel, G. (1968). *Naturwissenschaften* **55**, 84.
Merrifield, R. B., Stewart, J. M., and Jernberg, N. (1966). *Anal. Chem.* **38**, 1905.
Mross, G. R., and Doolittle, R. (1970). Personal communication.
Pisano, J. J., and Bronzert, T. J. (1969). *J. Biol. Chem.* **244**, 5597.
Sauer, R., Niall, H. D., and Potts, J. T., Jr. (1970). *Fed. Proc., Fed. Amer. Soc. Exp. Biol.* **29**, 728.
Schellenberger, A., Jeschkeit, H., Henkel, R., and Lehman, H. (1967). *Z. Chem.* **7**, 191.
Schroeder, W. A. (1967). *Methods Enzymol.* **11**, 445.
Schultz, J. (1967). *Methods Enzymol.* **11**, 255.
Stark, G. R. (1970). *Advan. Protein Chem.* **24**, 261.
Waterfield, M., and Haber, E. (1970). *Biochemistry* **9**, 832.

CHAPTER 5 Preparation of Cross-Linked Polystyrenes and Their Derivatives for Use as Solid Supports or Insoluble Reagents

JAMES A. PATTERSON

I.	Introduction	189
II.	Types of Polystyrene and Their Properties	190
	A. Soluble and Gel Polystyrenes	190
	B. Cross-Linked Polystyrenes	192
III.	The Preparation of Spherical Styrene-Divinylbenzene Copolymers	195
	A. General Considerations and Methodology	195
	B. Polymerization of Substituted Styrenes	200
IV.	Chemical Reactions of Styrene-Divinylbenzene Copolymers	201
	A. Sulfonation	201
	B. Chloromethylation	204
	C. Amination of Chloromethyl Polystyrene	206
	D. Properties of Ion-Exchange Polystyrene Resins	208
	E. Nitration of Polystyrene and Reduction of Nitropolystyrene	211
	References	212

I. INTRODUCTION

Beads of styrene-divinylbenzene copolymers have been used extensively for solid phase peptide synthesis (Merrifield and Marshall, Chapter 3) and as solid phase reagents or supports in the sequential degradation of peptides from the N-terminus (Stark, Chapter 4). In addition, polystyrene beads have been used as supports for the synthesis of polydeoxynucleotides (Letsinger and Mahadevan, 1966) and for the preparation of some insoluble derivatives of antigens (Goldman, Goldstein, and Katchalski, Chapter 1; Lindsey, 1970). It has usually been recognized that the physical and chemical properties of styrene-divinylbenzene copolymers profoundly affect their utility, but there has not yet been a review (from the point of view of the biochemist who wishes to synthesize a solid support or an insoluble reagent) of the way in which these properties vary with the conditions used for the synthesis of the polymers.

Some pertinent questions are (1) What is the molecular nature of the polymer? (2) How much is it cross-linked, and by what chemical groups? (3) How stable is the polymer matrix? (4) How accessible are the reactive sites? (5) What side reactions with polymer end groups are possible? (6) What impurities are incorporated within, or occluded to the resin matrix? (7) How homogeneous is the distribution of cross-linking within a polymer bead, and among different beads? (8) How homogeneous is the size distribution of the beads?

In this chapter, the preparation of styrene-divinylbenzene copolymers will be considered in the light of these questions. Similar general considerations apply to polymerization of other vinyl monomers such as acrylonitrile and substituted styrenes. Although the scope of this presentation is limited, I hope that an understanding of the chemistry of the simple cases discussed will enable biochemists to design better solid reagents in the future.

II. TYPES OF POLYSTYRENE AND THEIR PROPERTIES

A. Soluble and Gel Polystyrenes

The literature on polymerization of sytrene and similar unsaturated monomers has been extensively reviewed (Atlas and Mark, 1961; Rudd, 1960; McCormick, 1959; Kline, 1964). The scientific data necessary to predict the molecular weights of polymers, the extent of branching, and the relative rates of copolymerization for mixtures of monomers are available. In addition, classical chemistry has been shown to apply to aromatic polymer chains. An initial basic assumption for theoretical models was that the various molecular species are homogeneously distributed throughout the polymer.

Linear polymers can be solvated to form completely homogeneous dispersions in which the concentration of the polymer can be made to approach zero. In contrast, a "soluble" gel can be solvated homogeneously only to a limited extent, beyond which addition of more solvent will not increase the dilution of the polymer. The linear polymer can be swollen with solvent infinitely, whereas a soluble gel has a very large but finite swollen volume.

A transition from a soluble polymer to a gel accompanies an increase in branch formation or molecular weight. Unbranched and branched chains are illustrated in Fig. 1. Random branching of a linear polymer can be caused chemically, or by high energy radiation, heat, or ultrasonic radiation. Random transformation to the gel state by chain branching has been treated theoretically by Spiro *et al.* (1964). Radiation-induced chain branching has been obtained by Charlesby (1954, 1960) and also by Bovey (1958). Thermal and chemical chain transformations which increase the molecular weight or the degree of

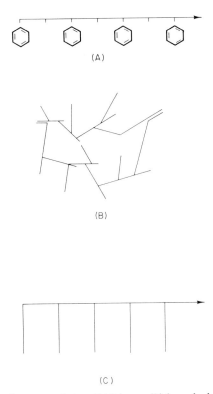

Fig. 1. Types of polystyrene chains. (A) Linear; (B) branched (unordered); (C) comb (ordered).

random branching have been discussed by Boundy and Boyer (1952). Controlled extension of the linear chain by branching is represented by the comb model (Fig. 1, C) in which increasing the molecular weight by branching does not produce a rapid transition from a soluble gel to an insoluble gel (Morton et al., 1962; Manson and Cragg, 1958; Gervasi and Gosnell, 1966; Altare et al., 1965; Zelinski and Wofford, 1965). Free rotation of the branched chains around the primary chain may be the characteristic which accounts for the increased relative solubility of the comb form of branched polystyrene.

Gels of random and ordered branched chain polymers which were initiated with linear polymers have been discussed by several authors (Flory, 1941, 1946, 1953; Stockmayer, 1952, 1953; Case, 1957; Thurmond and Zimm, 1952). Thurmond and Zimm studied the formation of gels by the inclusion of small amounts (0.0025%) of divinylbenzene (DVB). Mathematical models of gel polystyrene have been developed by Gordon (1962), Good (1963), and Dobson and Gordon (1965), in which the degree of rotational freedom and molecular weight determine physical and chemical characteristics of gels. Flory and Rhener

(1943) and Flory (1950) characterized swollen gels from the standpoint of matrix energy. The expression used for the degree of swelling of the gel has practical applications in indicating which solvents can be used with synthetic and natural polymers.

B. Cross-Linked Polystyrenes

The two most common geometric forms of cross-linked polystyrene are spherical beads and sheets. Formation of spherical beads is accomplished by suspension or emulsion polymerization; the size of the bead is determined by the technique used. Suspension polymerization is used for beads 1 to 3000 μ in diameter, and emulsion polymerization produces beads 0.02 to 1 μ in diameter. In either of these techniques, cross-links are formed by DVB, included during the polymerization. Other possibilities for cross-linking are chain entanglement, and cross-linking during subsequent chemical reaction of the beads, as occurs during chloromethylation (see below).

The sheet form of cross-linked polystyrene can also be subdivided into two classes: films and membranes. A film is considered to be from 0.005 for 2 μ thick, and a membrane to be from 2 μ to 1 mm thick. In general, films and membranes are cross-linked by the inclusion of DVB during the polymerization if they are to be subsequently converted to a hydrophilic form. Gregor (1966) has discussed the preparation and characterization of polymer films. For example, films can be formed from monomers (like the formation of divinylbenzene on oil surfaces) by electron bombardment. Other techniques of film polymerization are photolytic, gaseous discharge, and gas deposition. Cross-linking of membranes can be accomplished after polymerization by chemical reactions which result in the formation of polyfunctional sites, some of which can then be used for the formation of cross-links.

The rest of the discussion in this section will be limited to the bead form of polystyrene. Cross-linking by incorporation of a polyfunctional monomer into a growing chain of monofunctional monomer is controlled by the polymerization rate of each of each species, and by the ratio of their molar concentrations. Chain entanglement as an unstable source of polymer cross-linking must also be considered. The formation of styrene-DVB copolymers has been treated by Mayo and Lewis (1944). This treatment can be extended by applying the kinetic data for styrene and DVB found in "Styrene—Its Polymers, Copolymers and Derivatives" (Boundy and Boyer, 1952). By assuming a homogeneous mixture of the monomers and catalyst and equal free radical condensation activities for both types of vinyl groups, a model in which the polymer is homogeneous is obtained (Bauman and Eichorn, 1947). In this model, cross-linking is expressed as the initial percentage of DVB in the mixture of monomers, and each cross-link is assumed to appear with statistical regularity along the chain (Fig. 2). The model is thus seen to be an extension of the models for soluble polymers and gels.

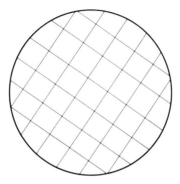

Fig. 2. Homogeneous cross-linked polymer model (fish net).

Chain entanglement is a different mode of cross-linking. Control of the degree of this low energy, unstable mode has been studied by Millar (1960) and Millar et al. (1962). These workers evaluated the effect of initiator concentration. Wiley et al. (1964) extended the work of Millar and his co-workers by studying the effects of heat and solvents on the degree of chain entanglement.

In the production of styrene-DVB copolymer beads by the suspension method, Pepper et al. (1953) noted the possibility that individual beads from a single reaction mixture might have different degrees of cross-linking (Fig. 3).

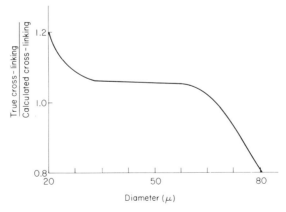

Fig. 3. Bead size as a function of cross-linking in a single polymerization.

Pepper suspected that the variation in cross-linking was due to inhomogeneous distribution of peroxide and to the different solubility of polymers in the monomers. Freeman *et al.* (1965) verified that there was a 2 to 15% deviation in cross-linking in commercially produced polymer beads. Freeman (1966) also noted that in highly cross-linked beads a deviation in cross-linking with respect to bead size was apparent where large beads from a single reaction mixture had a lower degree of cross-linking than small beads from the same mixture.

Spherical matrices of styrene-DVB copolymers can be represented by one or a combination of the models illustrated in Fig. 4. In model A, there is a random

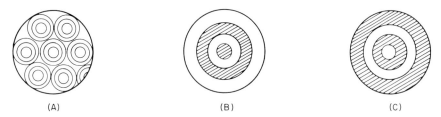

Fig. 4. Models of cross-linked copolymer beads. For details see text.

distribution of sites for initiating chain growth with two monomers of different polymerization rates. The rate of polymerization and molar concentration of each species determines its concentration gradient in concentric spheres during the three dimensional growth of the matrix. The center of each sphere represents the initiation site, and there is a gradient of cross-linking within each sphere. In styrene-DVB mixtures, the DVB polymerizes faster. Hence, the cross-linking at the initiation site will be higher than that in the terminal region of the concentric sphere of reaction. Therefore, each of the spheres within a bead will be connected by a zone of low cross-linking, forming a potential pore.

In model B, most of the initiating sites are at the center of the bead. This could, for example, result from inhibition of free-radical formation in the suspension medium. Here, the bead has a highly cross-linked center, with concentric

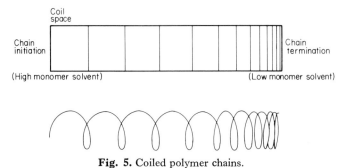

Fig. 5. Coiled polymer chains.

spheres of lower and lower degrees of cross-linking, out to the interface with the medium. Model C is the inverse of model B. The polymerization was initiated at the interface with the medium, and has progressed into the sphere of the monomer. The greatest degree of cross-linking is at the surface.

In each of the above models, the growing polymer is propagated as a helix in the presence of a high concentration of monomer. In the case of styrene-DVB copolymers, the monomers are a good solvent for the coiled growing polymer, and will produce a polymer in which a relatively loose coil is formed at initiation, with a tighter coil at termination (Fig. 5).

III. THE PREPARATION OF SPHERICAL STYRENE-DVB COPOLYMERS

A. General Considerations and Methodology

The use of underivatized (raw) styrene-DVB copolymer as a hydrophobic surface is usually complicated by the presence of the suspending or stabilizing agent at the surface (Frenkel et al., 1970). Although the degree of incorporation is low, these agents may perform an important function at the interface with solvent. Many times a research worker has unwittingly used an alumina interface when a polystyrene interface was desired. An inexplicable failure resulted. Impurities at the interface are very difficult to remove. Before using a particular polymer, therefore, one should know which suspension-stabilization system was used in its preparation. Thus, knowledge of the methodology used by the polymer chemist is of vital interest to the research worker who uses polystyrene resins as insoluble reagents, or as supports.

To understand how the models of Fig. 4 were derived, two currently used systems of polymer formation will be described in some detail, and a third will be mentioned. Suspension polymerization of styrene-DVB spheres occurs in a two-phase exothermic reaction, with a homogeneous mixture of monomer phase and medium phase. External heat is used to activate the decomposition of a source of free radicals which initiate the polymerization. The droplets of monomer are well dispersed in the medium by stirring. The medium consists of surface-active agents which lower the interfacial energy between the hydrocarbon and water phases, deionized, particle-free, double-distilled water, and surface-stabilizing agents which maintain the surface area developed. The nature of the medium is important in determining which of the models of Fig. 4 will be obtained. The monomer phase consists of DVB, monovinyls (styrene plus impurities in the DVB such as ethyl vinyl benzene; DVB is usually only about

55% pure), a catalyst (benzoyl peroxide), and a free radical inhibit to stabilize the monomer during storage, such as tertiary butyl catechol.

In a well-controlled, well-defined polymerization, all reactor surface must be inert (glass, for example), the temperature of the system must be controlled to $\pm 0.5°$, and oxygen and airborne particles must be excluded from the reaction. A system for controlled mixing must be available, and a means of recording the temperature of the reaction mixture as a function of time is also needed. Here a thermal feedback servo control can be usefully employed. In addition, the reaction mixture must be protected from light.

An inert surface, such as glass, is required to avoid interference with the formation of the free radicals which initiate polymerization. A few parts per million of iron, copper, or other metal salts can completely change the level and the rate of free radical formation. Temperature regulation is required for a controlled rate of free radical formation, for control of the viscosity of the system, and for control of the amount of monomer lost (or reversibly transferred) to the gas phase. Oxygen inhibits polymerization by trapping free radicals formed from the decomposition of the benzoyl peroxide. Airborne particles interfere with the control of the reaction by inhibiting activation, or by serving as nuclei. The mixing rate is another variable which determines the size of the spheres and the extent of their distribution according to size. As shown by Freeman (1966), the size and size distribution are associated with a variation in cross-linking from bead to bead.

A time-temperature record is required to determine whether the polymerization has followed the desired program. The mixed medium is first brought to a predetermined initial temperature, and the mixture of monomers is then added at a controlled rate. To avoid uncontrolled polymerization by free radicals formed from ultraviolet radiation, the entire system is protected from light.

DVB is one of the major contributors to polymer impurities. The highest purity of this monomer in practice is about 55%, with the remainder consisting of ethylvinylbenzenes and mono and diethylbenzenes. Styrene should be of high purity, because it determines principally the composition of the final polymer. The purity of the styrene also affects the chain length of the polymer, as impurities tend to be chain terminators. Tertiary butyl catechol in the styrene preserves the monomeric form during shipment and storage. The quantity of benzoyl peroxide catalyst used depends on how much of this preservative must be overcome before polymerization can be initiated. This amount of benzoyl peroxide is usually 0.25% to 2.0% of the total monomer weight. Formation of free radicals (Horner and Pohl, 1948) and interaction of the radicals with tertiary butyl catechol are illustrated in Figs. 6 and 7. The o-quinone or semiquinoine formed from tertiary butyl catechol can be trapped in the resin matrix and must be removed from the raw polymer.

Low concentrations of common salts, colloids, and organic material found in

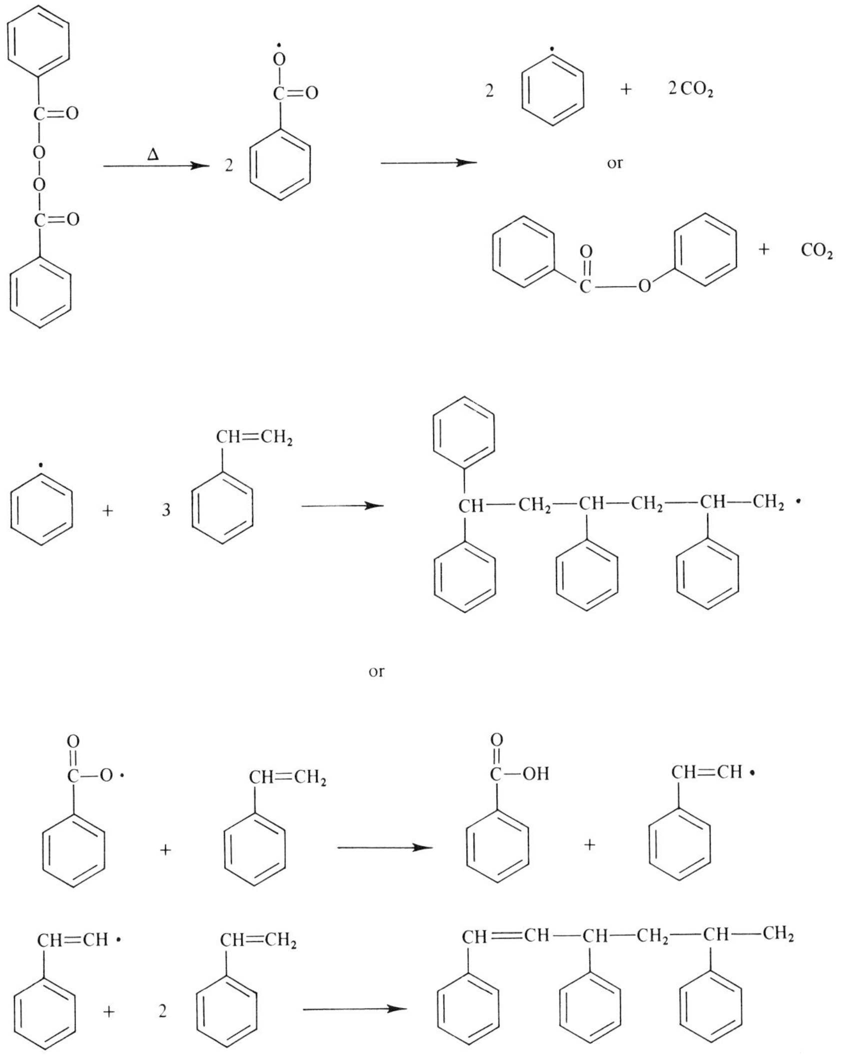

Fig. 6. Formation of free radicals (Horner and Pohl, 1948).

impure water will affect formation of free radicals from the benzoyl peroxide. Therefore, any reproducible suspension polymerization requires water of high purity.

The effect of medium, suspension, and stabilizing systems on the mode of polymerization (models A, B, and C of Fig. 4) will now be discussed. Commercially produced styrene-DVB copolymer spheres are of the model A type. The suspension and stabilizing agents are carboxymethyl cellulose (Dow Chemical

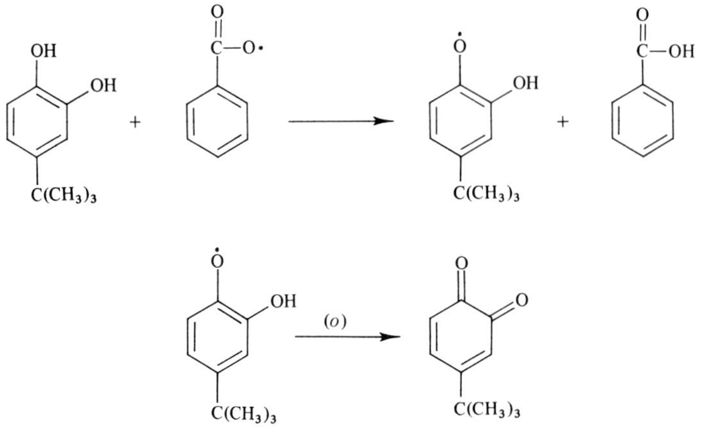

Fig. 7. Interaction of radicals with tertiary butyl catechol.

methocel), and hydrated colloidal alumina ($Al_2O_3 \cdot nH_2O$, National Aluminum Corp.), respectively. This suspension-stabilization system does not alter the randomness of free radical formation and initiation within the spheres of monomers. However, in this system, there are difficulties when the product is to be used where the majority of reaction will be at the surface. Alumina is incorporated into the interstitial spaces of the polymer surface (Fig. 8) and will continue to be leached from the polymer even after long periods of use. In general,

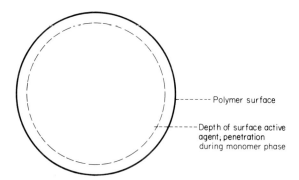

Fig. 8. Surface agent occlusion.

if the suspending and stabilizing agents are not thoroughly cleansed from the freshly prepared polymer before drying, a barrier to penetration can be baked onto the polymer surface. Then the conditions under which this barrier is overcome during subsequent chemical reaction may have to be more vigorous than the conditions necessary for pure polymer to react. This is seen in the so-called "penetration temperature" required in sulfonation of styrene-DVB for copolymer beads to form cation-exchange resins (see below).

As the rate of polymerization is decreased, the individual chains become longer, and minimum chain entanglement (pseudo cross-linking) occurs. However, such conditions cause increased incorporation of the stabilizing agent near the surface of the bead and greater agglomeration of the spheres as they pass through the sticky state. Lowering the concentration of DVB or benzoyl peroxide, or lowering the temperature decreases the polymerization rate. Lowering the sphere size by increasing the rate of mixing or the concentration of suspension-stabilizing agents will increase the polymerization rate to a point beyond which, determined by the number of free radicals that can form in each sphere, the pure model A (Fig. 4) will be lost, and an uncontrolled combination of model A and model B will result. This occurs at a mean particle diameter of about 100 μ and smaller. The extent to which the product resembles model B is a function of the concentration of free monomer remaining at the end of the free radical-catalyzed polymerization and at the onset of thermally initiated polymerization at the surface of the spheres.

By using Bentonite (colloidal particles of clay, about 1 μ in diameter) and gelatin in its isoelectric pH range, polymer spheres like those of model A (Fig. 4) can be produced, similar to the ones just described for the carboxymethyl cellulose-alumina system. By incorporating controlled amounts of metallic salts [e.g., $Fe(NO_3)_3$ or $Cu(NO_3)_2$], surface polymerization can be inhibited until some point in the polymerization program is reached, whereupon the salts can be precipitated by the addition of NH_4OH. Polymerization of the external portion of the spheres can then occur, resulting in a product like model B, with a highly cross-linked center and a less cross-linked exterior. Model B can also be produced by including in the medium oxygen or hydrogen peroxide, which lower the concentration of free radicals at the surface. Thus, uncontrolled mixing of air into a polymerization reaction can cause model B to result, even though another model is wanted. This is why most suspension and emulsion polymerizations are carried out under an inert blanket of nitrogen.

A spherical polymer of type C can be produced with the Bentonite-gelatin medium by initiating polymerization at the interface of the monomer sphere and the medium. Amino acid residues in gelatin act as activators, catalyzing decomposition of benzoyl peroxide. (Amines and transition elements such as Fe, Ni, and Co are classical activators of the formation of free radicals from organic peroxides.) When initiation of polymerization occurs at the interface, the degree of cross-linking will be higher there, since the rate of polymerization of DVB is greater than that of styrene.

There is a third suspension-stabilization system now commonly used which should be mentioned: Tergitol Anionic 7 (sodium heptadecyl sulfate, gum arabic). (Union Carbide Co.). As with the other suspension-stabilization systems, minor modifications can control the final mode of polymerization.

There are not many terminal groups in styrene-DVB copolymers; however, since chain termination occurs at the surface, the number of such groups there is relatively high. If only surface sites are to be utilized in some particular application, it may be important to consider the terminal groups as potential participants in the chemistry. The chain can terminate in a variety of chemical forms, as illustrated in Fig. 9. Reactive groups can also be produced by thermal and oxidative degradation of the polymer chain. See Boundy and Boyer (1952) for an extensive discussion of polymer degradation.

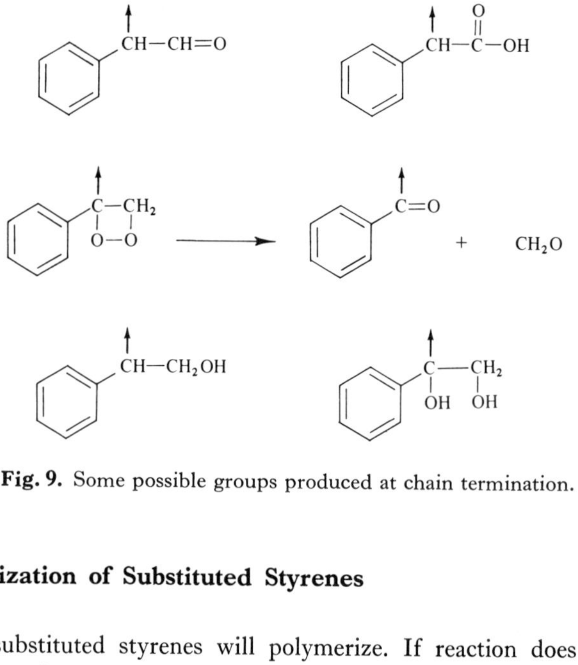

Fig. 9. Some possible groups produced at chain termination.

B. Polymerization of Substituted Styrenes

Not all substituted styrenes will polymerize. If reaction does occur, the polymerization of substituted styrenes with DVB as the cross-linking agent will result in polymer models different from those obtained with styrene under the same conditions, since substitution of the aromatic ring affects both the rate of polymerization and the lifetime of the free radical formed on the vinyl side chain. Thus, the mode of copolymerization with DVB may change. In designing substituted styrene-DVB copolymers, the substituted styrene should have a reaction rate similar to that of styrene, and it should also have similar polarity. Boundy and Boyer (1952), in Chapter 20 of their monograph, give tables of polymerization activity constants for a number of aromatic compounds. These values can be compared to those for styrene and DVB as a first approximation in designing conditions for the formation of substituted copolymers with DVB similar to the styrene-DVB copolymer models shown in Fig. 4.

IV. CHEMICAL REACTIONS OF STYRENE-DIVINYLBENZENE COPOLYMERS

A. Sulfonation

The methods which will be described now are commonly used in the production of cation-exchange resins for such different purposes as water softening and amino acid chromatography.

There are three general methods for sulfonating styrene-DVB copolymers: (1) H_2SO_4 alone, (2) H_2SO_4 plus perchloroethylene, and (3) chlorosulfonic acid plus methylene chloride. Before sulfonation, the raw polymer beads must be cleansed of as much suspension-stablization agents as possible, they must have had at least 12 hours of steam distillation to remove excess monomers and saturated aromatic impurities (from the DVB), and they must be free of moisture. The primary site of sulfonation is the position para to the ethylene side chain of the polymer, with a small amount of ortho substitution. The maximum amount of sulfonation is about 104%, based on the number of aromatic rings in the polystyrene.

1. H_2SO_4 Alone

The reactor usually is a covered glass vessel (polymer kettle) fitted with an overhead stirrer with provision for access through the reactor lid. The temperature of the reactor is continuously monitored and is usually controlled by a removable heating mantle. A good grade of H_2SO_4 is required for sulfonation; a concentration of between 96 and 105% (based on SO_3 content) is usual. The initiating temperature and heat capacity of the reaction mixture determine the approximate concentration of the acid. The polymer model (Fig. 4), degree of cross-linking, and bead size all have a bearing on the initiation temperature and exothermic slope.

The sulfonation reaction of a 7:1 molar mixture of H_2SO_4: polystyrene-DVB spheres is initiated at room temperature; this temperature is gradually increased until the penetration temperature is reached, at which point there is an inflection in the temperature-time curve due to the exothermic reaction (Fig. 10). At this point, the temperature is controlled at 5° above the penetration temperature, by cooling, and then allowing the heat of reaction to permit continued sulfonation. Sulfonation at high temperatures (for example, 145°) or with low concentrations of acid (92%), produces chain rupture due to oxidation, causing what is commonly called "color throw." The colored material is oxidized polymer and can continuously leach out over long periods. If a molecule under

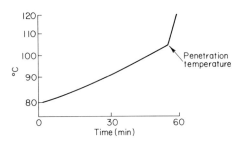

Fig. 10. Initiation or penetration temperature of 8% cross-linked polystyrene-DVB 600 μ mean diameter.

study is bound to one of these free polymer chains, it too may be lost by leaching. Oxidative rupture of the chains will also produce a variety of different polymer end groups, of which the type shown in Fig. 11 predominates.

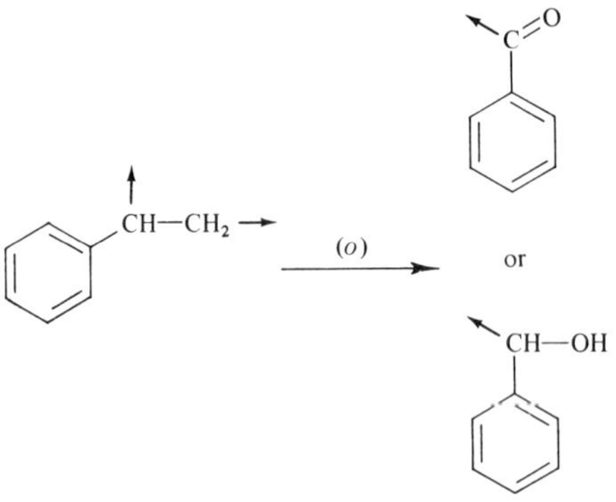

Fig. 11. Polymer end groups resulting from oxidative rupture of the chains.

2. H_2SO_4-Perchloroethylene

A reflux condenser is required to retain the perchloroethylene (B.P. 120°) in the reaction vessel. The molar ratio of sulfuric acid to polystyrene-DVB spheres is the same as that required for sulfonation with H_2SO_4 alone. Perchloroethylene (10–30% of the bulk volume of the polystyrene-DVB spheres) is added to the polymer at least 1 hour prior to reaction. The solvent expands the spheres and shifts the sulfonation initiation curve (Fig. 10) approximately 10° downward. Lowering the temperature reduces degradation of the polymer and minimizes formation of sulfone linkages (see below). After the controlled reaction, as followed by the exotherm, is terminated, the acid is diluted, the beads are washed with water, and the excess perchloroethylene is removed by steam distillation.

3. ClSO₃H-Methylene Chloride

The glass reactor used for the previous sulfonation is used, with a water cooled condenser added. Heat is not required, but a vented container for the dropwise addition of ClSO₃H is a necessity. Methylene chloride (twice the bulk volume of the polymer) is required to expand the spheres fully. The swelling is carried out in the reactor. The swollen mixture should be a liquid slurry; if it is not, additional methylene chloride should be added. After a half hour of mixing, ClSO₃H is slowly added. A molar ratio of 1.8 moles of ClSO₃H to 1.0 mole of polystyrene-DVB beads is required (Fig. 12). This method of sulfonation produces an increased number of sulfone cross-links (Fig. 13). At approximately

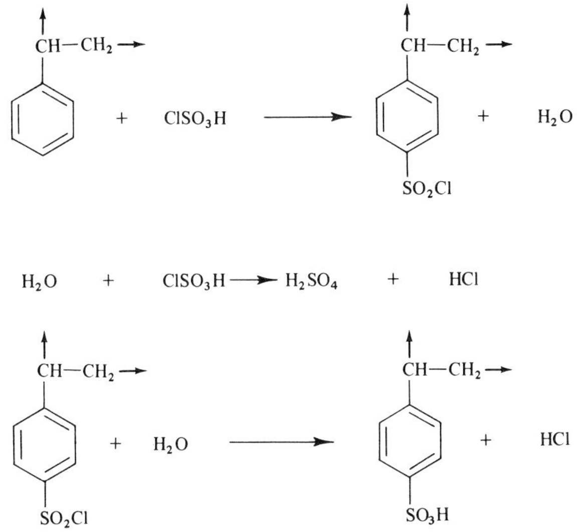

Fig. 12. Sulfonation of styrene-DVB copolymers with H_2SO_4 plus perchloroethylene.

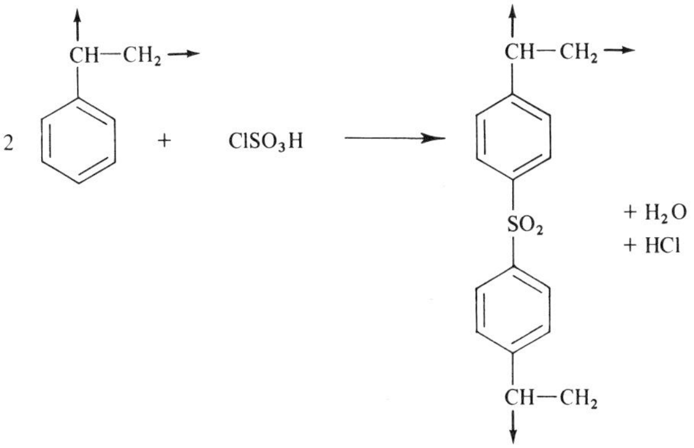

Fig. 13. Increased number of sulfone cross-links produced by sulfonation with H_2SO_4 plus perchloroethylene.

30% reaction, the lyophilic nature of the polymer changes; and the polymer agglomerates and disperses again only after subsequent dilution with water. After dilution, the lower layer of excess methylene chloride is decanted off, and the remaining methylene chloride in the reacted spheres is removed by steam distillation.

B. Chloromethylation

Production methods for the common anion-exchange resins and the chloromethylated intermediate result in a polymer with a matrix of the type illustrated in Fig. 4B. The polymer at the interface is initially more cross-linked than the polymer in the interior, but NaOH in the production process causes hydrolysis of methylene bridges preferentially at the surface to give a final product with less cross-linking at the surface, with a zone of more highly cross-linked polymer, and a less cross-linked center.

Chloromethylation of the free aromatic rings of polystyrene-DVB with chloromethyl methyl ether is a Friedel-Crafts reaction, with orientation para to the ethylene chain. In addition to chloromethylation of the aromatic rings, methylene bridging will occur within and between chains (Fig. 14). Commercial anion-exchange resins and chloromethlyated intermediates utilize both chemical reactions of Fig. 14. Methylene bridging is used for 50 to 80% of the reported nominal 8% cross-linking, i.e., the starting polymer is cross-linked with 1 to 2% DVB. The degree of methylene bridging is controlled by the purity of the chloromethyl methyl ether (CMME), the volume of CMME used relative to the volume of polymer spheres, the amount of initial DVB cross-linking, the purity of the anhydrous Friedel-Crafts reagent, and the amount of Friedel-Crafts reagent used. Freshly fused Friedel-Crafts catalysts and freshly distilled CMME are recommended, since these reagents deteriorate on storage.

Not only is CMME involved in the chemical reaction of chloromethylation, but it is also the swelling agent for the polymer. If any decomposition products of CMME are present, solvation is impaired and the matrix will be contracted. The contracted matrix has a lower reaction rate. A higher proportion of the reacted sites will give rise to bridging. Water decomposes the Friedel-Crafts reagent [$AlCl_3 + 3H_2O \rightarrow Al(OH)_3 + 3HCl$], destroying its catalytic activity and promoting decomposition of the CMME.

CMME is added in excess to the polystyrene-DVB beads in a moisture-free, stirred glass reactor. The polymer is allowed to expand completely, adding more CMME as necessary to maintain a slurry. This swelling period at ambient temperature is approximately half an hour. Either of the two commonly used Friedel-Crafts reagents ($AlCl_3$ or $ZnCl_2$) is then suspended in a portion of CMME and then added to the swollen polystyrene-DVB slurry. The molar

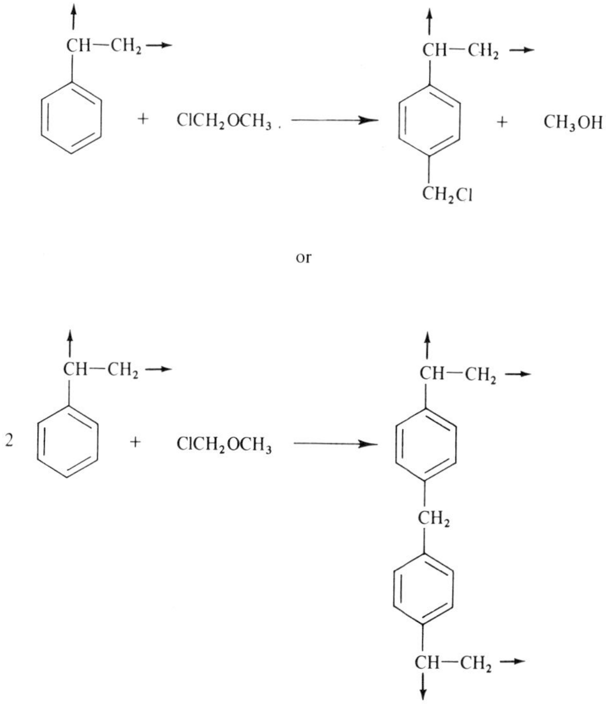

Fig. 14. Chloromethylation of the aromatic rings and methylene bridging within and between chains.

relationship of polystyrene-DVB to Friedel-Crafts reagent is 1.0 to 1.25; excess Friedel-Crafts reagent will not interfere with the reaction. A reaction time of 2 to 6 hours is then allowed. The Friedel-Crafts reagent is not a true catalyst, but forms a reactive complex with the CMME and polymer. Some Friedel-Crafts reagent does remain in the polymer matrix. There is at least one other matrix effect as a result of chloromethylation: the relaxing of chain entanglement pseudo cross-links. Having a low content of DVB, copolymers with this relaxation of chain entanglement allow those polymer chains which have no chemical cross-link to be leached from the matrix. These leachable polymers are called "linear polymers," "color throw," or "soluble polymer solids." In cases where an appreciable amount of linear polymer is released from the matrix, the whole reaction mixture can become a gel-like mass until the linear material has been extracted. Under these circumstances, the original degree of cross-linking is lost; i.e., a 1% cross-linked polymer with sufficient linear material leached from the matrix can become about 2% cross-linked.

Extraction of excess CMME and Friedel-Crafts reagent is accomplished by pouring the reaction mass over ice-cold 1 M HCl and washing the polymer solid on a filter with cold 1 M HCl. If the chloromethyl intermediate is to be used, for example, for peptide synthesis, the product should be washed until the effluent is free of the Friedel-Crafts reagent. (This can be evaluated by atomic absorption spectroscopy.) In addition to the Friedel-Crafts reaction for the incorporation of sites reactive toward amines on the aromatic ring, there is at least one other commonly used method, the reaction of polystyrene spheres which are not cross-linked with a complex mixture of $ClSO_3H$ and formaldehyde. In this reaction, a cross-linked polymer matrix of the type illustrated in Fig. 4B is produced, along with sites for amine substitution. The chemical nature of this reaction has not been fully disclosed.

The chloromethylated intermediates are used in the synthesis of type I (strong base) and type II (intermediate base) anion-exchange resins, and also in solid phase peptide synthesis (Merrifield and Marshall, Chapter 3).

C. Amination of Chloromethyl Polystyrene

Type I anion-exchange resins are produced by reaction of trimethylamine (TMA) with the chloromethylated polystyrene-DVB polymer sphere, and type II anion-exchange resins are made in a similar reaction with dimethylethanolamine (DMEA). Production of weak base ion-exchange resins by the reaction of secondary and primary amines requires somewhat more drastic conditions than those needed for the strong and intermediate base resins which are produced at ambient temperatures. For example, the weak base resin produced with triethylene tetramine requires activation at a temperature above 100°. The reaction of TMA with chloromethylated polystyrene-DVB will serve as a model for both strong and intermediate base-anion exchange resins. To avoid alkaline hydrolysis of the chloromethyl site (Fig. 15), the intermediate is suspended in a liquid such

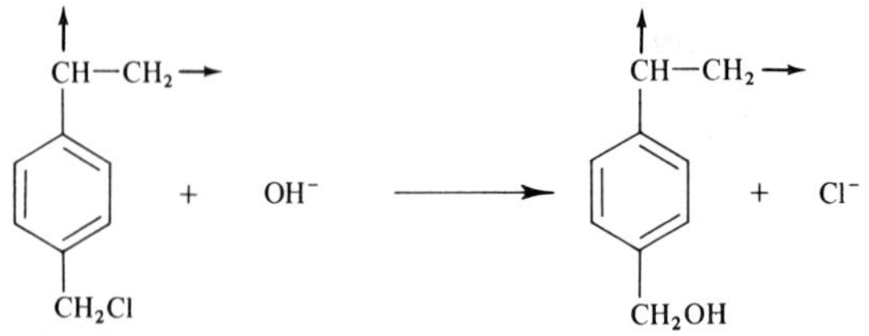

Fig. 15. Suspension of intermediate base-anion exchange resin in isopropanol to avoid alkaline hydrolysis of the chloromethyl site.

as isopropanol. TMA (gas) is added to this stirred suspension. After a molar ratio of 2:1 TMA:polystyrene-DVB has been added, 6 M NaOH (1:1 molar ratio to polystyrene-DVB) is added slowly to yield the basic form of the resin

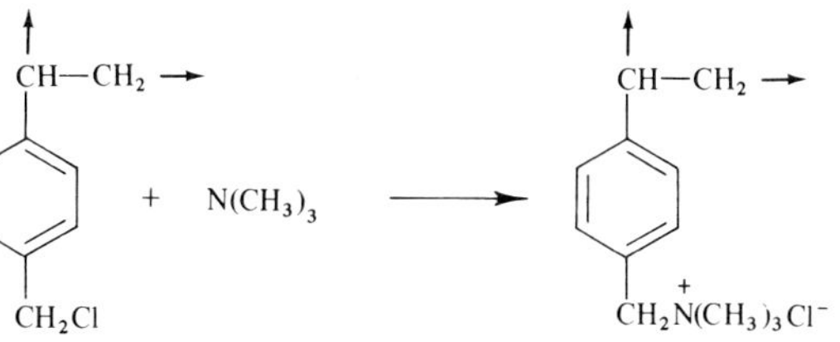

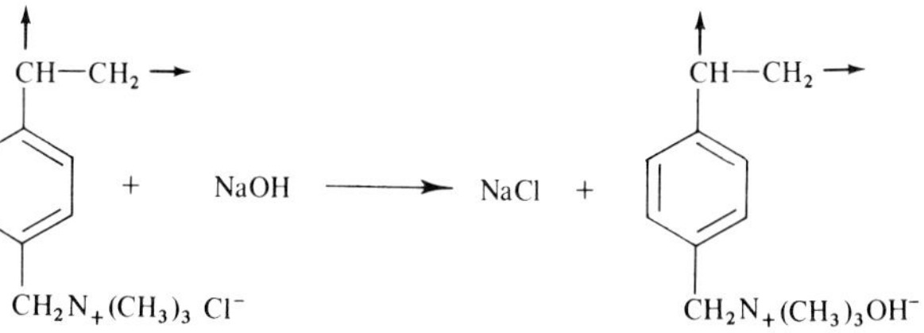

Fig. 16. Addition of NaOH to suspension to yield basic form of the resin.

(Fig. 16). The polymer is maintained in maximum expanded form in isopropanol in the free base form. Small amounts of HCl resulting from hydrolysis of the Friedel-Crafts reagent by small amounts of water (even the best purification procedure does not completely eliminate all of the Friedel-Crafts reagent) will form salts with the amino groups of the polymer which will contract the polymer matrix and terminate the amination reaction. Thus, the NaOH neutralization has a twofold function.

The aminated product is then filtered and washed with a mixture of HCl (0.5 M) and sodium chloride (2 M) on a filter until all free TMA and isopropanol have been removed. Then, in HCl–NaCl solvent, the product is steam distilled to remove traces of solvent.

On the basis of monomer chemistry, one might predict that the strong base quaternary salt would be the only product. However, the final product generally has a mixture of strong base quaternary sites and weak base tertiary sites in ratios from 80:70 to 30:70 on an equivalent basis. While the mechanism has not

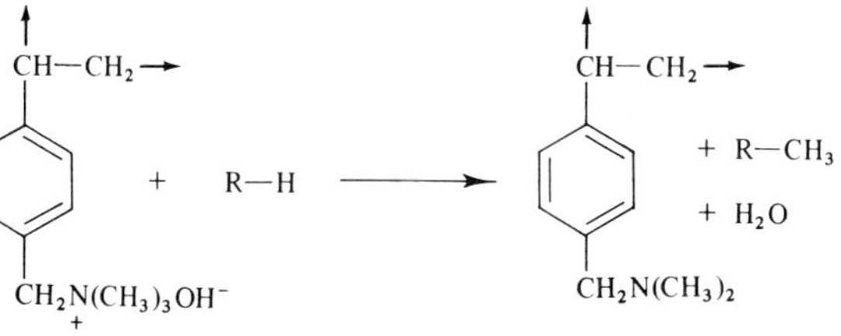

Fig. 17. Functioning of the quaternary base as an alkylating agent.

been fully studied, a reasonable explanation of the origin of the weak base-exchange sites is that the quaternary base can function as an alkylating agent (Fig. 17).

Sites for methylation are also available in the aliphatic chain of the polymer and in impurities which were incorporated in the resin matrix before chloromethylation.

D. Properties of Ion-Exchange Polystyrene Resins

Because the cationic and anionic derivatives of cross-linked spherical polystyrene-DVB beads, which form hydrophilic matrices for ion exchange in aqueous solution, are commonly used, they have been studied most extensively. It is wise to consider the behavior of these charged derivatives for some prospective use in which the functional groups of the resin will not be charged, since the behavior of the charged polystyrenes reflects general aspects of the structure of the polymer. As discussed above, the chemical reactions used to produce the anionic or cationic polystyrene-DVB resins produce different matrices. For example, in an 8% cross-linked cation-exchange resin, almost all of the cross-links are derived from DVB. However, in an 8% cross-linked anion-exchange resin, only about 25% of the cross-links are due to DVB; the rest are methylene bridges which are formed during the chloromethylation reaction.

Boyd and Soldano (1953) compared the swelling of lyophilic underivatized polystyrene-DVB beads with that of anion- and cation-exchange resins in various ionic forms. This comparison indicated a greater deviation in the cross-linking of the anion- and cation-exchange resins than in the underivatized polymer. Reichenberg and McCauley (1955) considered that three kinds of regions might exist in the same sulfonated polystyrene bead: (1) low cross-linked regions in which the full hydrated volume of the bound ions could be accommodated and which would be expected to have low selectivity for different ions; (2) intermediate cross-linked regions, where the size of the hydrated ion and the energy of its interaction with the binding site would be the bases of selectivity; (3) highly cross-linked regions, where the energy of stripping off water of hydration from the ions would be the major factor in ion selectivity. The rate of ion exchange would be higher for the low cross-linked zones and would decrease with increasing cross-linking. From the theoretical standpoint, this was a sharp deviation from the homogeneous gel model to one of zone heterogeneity. Freeman (1966), in his microscopic statistical study of individual beads, demonstrated that heterogeneity exists among individual beads from a single polymerization. The larger beads have less cross-linking, and the greater the difference, the greater the cross-linking.

Spherical cation-exchange resins of sulfonated polystyrene-DVB have been the most commonly investigated. To account for deviations from the behavior predicted by the Donnan equilibrium expression, Davies and Yeoman (1953) postulated that the system which they studied (8% cross-linked resin) had 5% of the total resin volume as voids filled with external electrolyte. Their model had valid relationships for a limited range of external electrolyte concentrations. For low cross-linked resins, Katchalsky (1954) reviewed the field and supported a model for the matrix which was similar to a linear polyelectrolye. A theoretical model of concentric charged zones of the spherical resin, maintained in balance by elastic forces, was prepared by Lazare et al. (1956). This model did not predict the observed activity coefficient for the resin phase. Impaired site function was given as the reason for the large (50%) deviation from the Donnan law observed by Nelson and Kraus (1958) for a 16% cross-linked cation-exchange resin. Good correlation was observed for a 1% cross-linked resin. Matrix heterogeneity was shown by Grubhofer (1959) with the aid of electron microscopy of the dehydrated, contracted matrix of a section of an 8% cross-linked cation-exchange resin. Nagasawa and Rice (1960) also compared heterogeneous and homogeneous models and demonstrated the contribution of the density of charge sites to an ion-exchange reaction. In 1961, Rice and Nagasawa extended their theoretical model for the resin to include electrostatic binding forces and spatial orientation of charged sites. In correlating electrostatic dielectric constants, the resin was found to be lower than free water in a cation-exchange resin of 5 meq/g capacity. The dehydrated exchange sites were found to have an approximate volume of 300 $Å^3$ and a radial separation of 7 Å. They also predicted greater chain entanglement (pseudo cross-linking) for resins of higher DVB content. Griessbach et al. (1961) postulated diffusion channels of unbound water in the resin matrix, where ion selectivity was based on site charge density and the dielectric constant of the matrix. Free electrolyte uptake was based on an electrical double layer model by Shone (1962). This model gave charged sites on 60 Å concentric radii; at the primary surface, the sites were 9 Å point to point with a charge density of 2.25×10^{-10} Faraday/cm^2. Further, Shone's model gave a heterogeneously charged resin phase which does not obey the Donnan law. Glueckauf and Watts (1962) gave an excellent summation of the status of resin phase heterogeneity and deviations from the behavior predicted from the Donnan law. In addition, they pinpointed areas of conflict requiring further investigation. A number of investigators (Dinius et al., 1963; Reichenberg and Lawrenson, 1963; Dinius and Choppin, 1964) extended Gordon's (1962) concept of bonding energy of the exchange site to show the possibility of a spectrum of bonding energies from Van der Waals, to hydrogen bonding, to relatively strong ionic bonds. Goldring (1962) precipitated silver chloride in the potential free space of a fully hydrated 8% cross-linked cation exchange resin. These loaded matrices were then sectioned and dehydrated for electron microscopy. Particle sizes of silver chloride were found to be as large as 1000 Å, with a mean in the 400–600 Å range. Past calculations of transport pore size generally agreed that the 10 Å range fit the observed data. This author suggests that a network of interconnecting capillaries of 10 Å exists between the larger voids which are observed (400–600 Å) and that these small capillaries function as the limiting barrier in ion transport.

The treatment of anion-exchange resins has not been as extensive as that of sulfonated cation-exchange resins. Hale et al. (1953) attributed increased pseudo cross-linking over that of the DVB to polymer chain entanglement during the chloromethylation step of polymer activation. Upon reacting the chloromethylated polymer with TMA, they found that approximately 20% of the chain entanglement cross-linking was lost. To account for deviation of ion exchange behavior from the Donnan law, Kraus and Moore (1953) attributed their observed data to the formation of tertiary and secondary amine sites during the reaction with TMA to form a quaternary strong base resin (see the earlier discussion on the preparation of anion-exchange resins). Freeman (1960) accounted for his observed deviation from Donnan theory at high external electrolyte concentrations by invoking chemical impurities in the matrix and external electrolyte which had diffused into the internal matrix. The deviations were corrected for by two arbitrary constants in his expression of the Donnan equilibrium expression. The difference in transport function of anions and cations with respect to hydration and hydrational radii energy levels was treated by Chu et al. (1962). In addition to the hydrational concept of ionic selectivity of exchange, the hydration of the immobile resin phase was treated. Finally, functions for deviation from the Donnan expression due to charge site density and chemical heterogeneity (quaternary, tertiary, or secondary amine sites) were combined into a complete equilibrium expression. Methylene bridging during the Friedel-Crafts reaction was considered by Anderson (1964).

Variables in the ion-exchange matrix important for use of the resin in amino acid analysis were discussed by Hamilton (1963). Resins for use in conjunction with accelerated methods of amino acid analysis were then developed, applying the polymer model of Fig. 4B (Benson and Patterson, 1965a,b; Benson et al., 1966). Green and Anderson (1965), in working out a system for chromatography of carbohydrates on anion-exchange resins, encountered difficulties in finding supplies of resin which would give reproducible results. Through design of a specific resin matrix, controlled specifications were established for an improved chromatographic systems. (Ohms et al., 1967). Scott et al. (1967) extended the applications of liquid chromatography on the same resin matrix.

Although ionic membranes will not be treated here, it should be emphasized that such membranes will compare in a limited way to a very low cross-linked sperical ion-exchange resin.

E. Nitration of Polystyrene and Reduction of Nitropolystyrene

Aromatic derivatives of the types R-NN, R-OH, R-NCO, and R-NCS are useful in biological applications and can be synthesized from amino polystyrene. The chemistry required to aminate low cross-linked (0.25 to 1%) polystyrene-DVB spheres is therefore of interest. The first step is nitration of the unsubstituted resin. The oxidative events which occur during nitration are polymer degradation and relaxation of chain entanglement to produce linear polymer. To avoid agglomeration of the spheres by excess linear polymer after nitration, an initial leaching of the polymer should be performed. The method I have used is (1) leach with 6 M HCl (1 volume of beads, 2 volumes of acid, 30 minutes at 80°), (2) filter and wash with water, (3) leach with 2 M NaOH (1 volume of beads, 2 volumes of NaOH, 30 minutes at 80°), (4) filter and wash with water, (5) air dry the beads, (6) leach them with acetone, (7) filter and wash with fresh acetone, and (8) leach with methylene chloride. Then completely swell the beads with methylene chloride, with a 15% excess as solvent for the linear polymer. Allow the extraction to proceed for 24 hours, then filter. Replace the methylene chloride and repeat the leaching procedure until an evaporated sample of the solvent shows no visible residue. (9) Filter the beads, (10) store them in methylene chloride, or use the swollen beads directly in the nitration reaction.

The conditions I have used for nitration of low cross-linked polystyrene-DVB beads are those suggested by Zenftman and McLean (1949), which minimize degradation of the polymer chains. The variables controlled were time, temperature, and the concentrations of HNO_3, H_2SO_4, and water. Although the optimum conditions given by Zenftman and McLean have been followed and verified, large amounts (up to 15% by weight of the total polymer for 0.25% polystyrene-DVB) of linear soluble nitrated polymer will be produced by oxidative degradation.

Prior to the reduction step, all linear nitrated material must be extracted from the resin with warm dimethylformamide on a filter. This extraction is continued until the dimethylformamide wash is colorless. The beads should be a transparent light brown with no opaque centers.

In the reduction of nitrated polystyrene-DVB with $SnCl_2$ (Fig. 18), the water of reaction, dilution, and hydration will control the degree and rate of reaction.

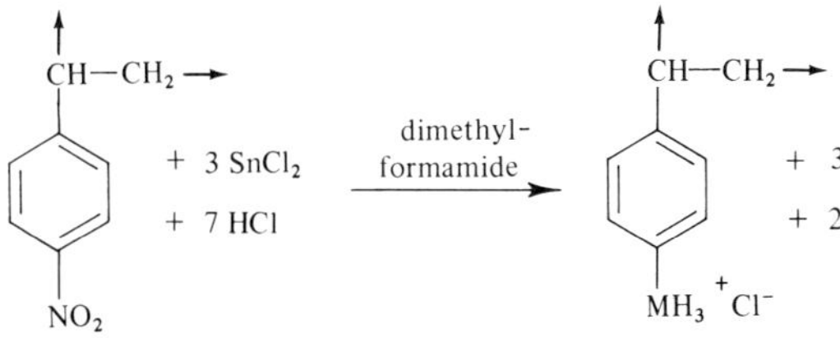

Fig. 18. Reduction of nitrated polystyrene-DVB with $SnCl_2$.

Water in a mixed solvent with dimethylformamide will also affect the swelling of the nitrated and reduced polymer. Following reduction, complete leaching of tin salts from the polymer is essential. This can be accomplished with warm 2 M HCl and dimethylformamide (50 : 50 mixture). After leaching, the amine polymer should be used immediately or stored under refrigeration as a heavy slurry in dimethylformamide. When dry, the amine polymer reacts to form a black product in air and sunlight; the nature of this product is unknown. The conversion of amino polystyrene to NCS polystyrene has been discussed by Stark (Chapter 4); references to other reactions can be found in the articles by Goldman, Goldstein, and Katchalski (Chapter 1) and Lindsey (1970).

Acknowledgment

Thanks are due to Dr. Stark without whose persistence this chapter would be unwritten and Karen Shore without whose assistance it would never have been assembled.

References

Altare, T., Jr., Wyman, D. P., Allen, V. R. and Meyersen, K. (1965). *J. Polym Sci., Part A* **3**, 4131.
Anderson, R. E. (1964). *Ind. Eng. Chem., Prod. Res. Develop.* **3**, 85.
Atlas, A. M. and Mark, H. F. (1961). "Report on Molecular Weight and Measurements of Standard Polystyrene Samples." Commission on Macromolecules of IUPAC, Montreal.
Bauman W. C. and Eichorn, J. (1947). *J. Amer. Chem. Soc.* **69**, 2830.
Benson, J. V. and Patterson, J. A. (1965a). *Anal. Chem.* **37**, 1108.
Benson, J. V. and Patterson, J. A. (1965b). *Anal. Biochem.* **13**, 265.
Benson, J. V. Jones, R. T., Cormick, J., and Patterson, J. A. (1966). *Anal. Biochem.* **16**, 91.
Boundy, R. H. and Boyer, R. F. (1952). "Styrene, It's Polymers and Derivatives." Reinhold, Pub. Corp., New York.
Bovey, F. A. (1958). *Polym. Rev.* **1**, 61–65 and 73–90.
Boyd, G. E. and Soldano, B. A. (1953). *J. Amer. Chem. Soc.* **75**, 6019.
Case, L. C. (1957). *J. Polym. Sci.* **26**, 333.
Charlesby, A. (1954). *Proc. Roy. Soc., (Ser.), A* **222**, 60 and ibid 542.
Charlesby, A. (1960). "Atomic Radiation and Polymers," Pergamon Press, Oxford.
Chu, B., Whitney, D. C., and Diamond, R. M. (1962). *J. Inorg. Nucl. Chem.* **24**, 1405.
Davies, C. W. and Yeoman, G. D. (1953). *Trans. Faraday Soc.* **49**, 968.
Dinius, R. H. and Choppin, G. R. (1964). *J. Phys. Chem.* **68**, 425.
Dinius, R. H., Emerson, M. T., and Choppin, G. R. (1963). *J. Phys. Chem.* **67**, 1178.
Dobson, G. R. and Gordon, M. (1965). *J. Chem. Phys.*, **43**, 705.
Flory, P. J. (1941). *J. Amer. Chem. Soc.* **63**, 3083.
Flory, P. J. (1946). *Chem. Rev.*, **39**, 137.
Flory, P. J. (1950). *J. Chem. Phys.* **18**, 103.
Flory, P. J. (1953). "Principles of Polymer Chemistry," Chapter X. Cormell Univ. Press, Ithaca, New York.
Flory, P. J. and Rhener, J., Jr. (1943). *J. Chem. Phys.* **11**, 521.

Freeman, D. H. (1960). *J. Phys. Chem.* **64**, 1048.
Freeman, D. H. (1966). *In* "Ion Exchange" (J. A. Marinsky, ed.), p. 197. Marcel Dekker, New York.
Freeman, D. H., Patel, V. C., and Smith, M. E. (1965). *J. Polym. Sci., Part A* **3**, 2893.
Frenkel, J. K., Dubey, J. P., and Miller, N. L. (1970). *Science* **167**, 893.
Gervasi, J. A. and Gosnell, A. B. (1966). *J. Polym. Sci., Part A* **4**, 1391 and 1401.
Glueckauf, E. and Watts, R. E. (1962). *Proc. Roy. Soc., (Ser.) A* **268**, 339.
Goldman, R., Goldstein, L., and Katchalski, E. (1971), this volume.
Goldring, L. S. (1962). *Abstr. 142nd Meet., Amer. Chem. Soc.* Atlantic City, New Jersey.
Good, I. J. (1963). *Proc. Roy. Soc., (Ser.) A* **272**, 54.
Gordon, M. (1962). *Proc. Roy. Soc. (Ser.) A* **268**, 240.
Green, J. G. and Anderson, N. G. (1965). *Fed. Proc., Fed. Amer. Soc. Exp. Biol.* **24**, 606.
Gregor, L. V. (1966). *Phys. Thin Films*, **3**, 131.
Griessbach, R., Gartner, K., and Anton, E. (1961). *Kolloid-Z.* **175**, 123.
Grubhofer, N. (1959). *Makromol. Chem.* **30**, 96.
Hale, D. K., Packham, D. I., and Pepper, K. W. (1953). *J. Chem. Soc. London*, p. 844.
Hamilton, P. B. (1963). *Anal. Chem.* **35**, 2055.
Horner, L. and Pohl, H. (1948). *Justus Liebigs Ann. Chem.* **559**, 48.
Katchalsky, A. (1954). *Progr. Biophys. Biophys. Chem.* **4**, 1.
Kline, G. M. (1964). *Mod. Plast.* **41**, 182.
Kraus, K. A. and Moore, G. E. (1953). *J. Amer. Chem. Soc.* **75**, 1457.
Lazare, L., Sundheim, B. R., and Gregor, H. P. (1956). *J. Phys. Chem.* **60**, 641.
Letsinger, R. L., and Mahadevan, V. (1966). *J. Amer. Chem. Soc.*, **88**, 5319.
Lindsey, A. S. (1970). *J. Macromol. Sci., Rev. Macro. Chem.* **4**, 1.
McCormick, H. W. (1959). *J. Polym. Sci.* **36**, 341.
Manson, J. A. and Cragg, L. H. (1958). *J. Polym. Sci.* **33**, 193.
Mayo, F. R. and Lewis, F. M. (1944). *J. Amer. Chem. Soc.*, **66**, 1954.
Merrifield, G. R., and Marshall, R. B. (1971), this volume.
Millar, J. R. (1960). *J. Chem. Soc., London*, p. 1311.
Millar, J. R., Smith, D. G., and Marr, W. E. (1962). *J. Chem. Soc., London*, p. 1889.
Morton, M., Helminiak, T. E., Gadgary, S. D., and Bueche, F. B. (1962). *J. Polym. Sci.* **57**, 471.
Nagasawa, M. and Rice, S. A., (1960). *Amer. Chem. Soc.*, **82**, 5070.
Nelson, F. and Kraus, K. A. (1958). *J. Amer. Chem. Soc.* **80**, 14154.
Ohms, J. I., Zec, J., Benson, J. V., and Patterson, J. A. (1967). *Anal. Biochem.* **20**, 51.
Pepper, K. W., Paisley, H. M., and Young, M.A. (1953). *J. Chem. Soc., London*, p. 4097.
Reichenberg, D. and Lawrenson, I. J. (1963). *Trans. Faraday Soc.* **59**, 141.
Reichenberg, D. and McCauley, D. J. (1955). *J. Chem. Soc., London*, 2471.
Rice, S. N. and Nagasawa, M. (1961). "Polyelectrolyte Solutions," pp. 461ff. Academic Press, New York.
Rudd, J. F., (1960). *J. Polym. Sci.* **40**, 459.
Scott, C. D., Attrill, J. E., and Anderson, N. G. (1967). *Proc. Soc. Exp. Biol. Med.* **125**, 181.
Shone, M. G. T. (1962). *Trans. Faraday Soc.* **58**, 805.
Spiro, J. G., Goring, D. A., and Winkler, C. A. (1964), *J. Phys. Chem.* **68**, 323.
Stark, G. R. (1971), this volume.
Stockmayer, W. H. (1953). *J. Polym. Sci.* **2**, 69.
Stockmayer, W. H. (1953). *J. Polym. Sci.*, **11**, 424.
Thurmond, C. D. and Zimm, B. H. (1952). *J. Polym. Sci.*, **8**, 477.
Wiley, R. H., Allen, J. K., Change, S. P., Mussleman, K. E., and Venkatachalam, T. K. (1964). *J. Phys. Chem.* **68**, 1776.
Zelinski, P. and Woffard, C. F. (1965). *J. Polym Sci., Part A* **3**, 93.
Zenftman, H., and McLean, A., Brit. Pat. 616,453 (1949).

Author Index

Numbers in italics refer to the pages on which the complete references are listed.

A

Adams, J. B., Jr., 136, *164*
Adler, A. J., 83, 104, *107*
Akanuma, Y., 102, *107*
Akeson, W., 86, *109*
Albertson, P. A., *153*, 162
Albu-Weissenberg, M., 10, 13, 18, 22, 25, 28, 77
Aldrich, F. L., 5, 14, 24, *72*
Alexander, B., 10, 17, 28, 64, 65, *73*, *74*
Aliapoulios, M. A., 104, *107*
Allen, J. K., 193, *213*
Allen, V. R., 191, *212*
Altare, T., Jr., 191, *212*
Anderson, B. S., 5, *75*
Anderson, N. G., 210, *213*
Anderson, R. E., 210, *212*
Andersson, B., 105, *109*
Anfinsen, C. B., 11, 67, *74*, 79, 81, 84, 85, 87, 95, 98, 99, 100, 101, 102, *107*, *108*, *109*, 121, 122, 124, 161, *162*, *166*, *167*
Anton, E., 209, *213*
Arnold, W. N., 13, *73*
Arsenis, C., 83, 98, *107*
Ashoor, S. H., 13, *73*
Atkinson, B., 42, *73*
Atlas, A. M., 190, *212*
Atlas, D., 10, 11, 12, 19, 20, 24, 25, 27, 28, 29, *75*
Attrill, J. E., 210, *213*
Augenstein, L., 3, 5, *76*
Augustin, J., 185, *187*
Avrameas, S., 4, 25, *73*, 77
Axén, R., 8, 11, 17, 24, 29, *73*, *74*, 77, 84, 87, *107*, *109*, *162*, *167*, 187, *187*

B

Babcock, K. L., 4, *76*
Bachler, M. J., 4, *73*

Backer, T. A., *162*
Baker, B. R., 67, *73*
Baker, H., 28, 66, *76*
Baker, H. R., 86, *109*
Bar-Eli, A., 12, 13, 18, 24, 28, 29, 39, 40, *73*, 77
Barker, S. A., 7, *73*
Barnett, L. B., 5, *73*
Barshad, I., 4, *76*
Bass, L., 51, *73*
Bath, R. J., *163*
Baugh, C. M., 121, 141, 158, 160, *165*
Bauman, E. K., 5, 14, 24, 68, *73*
Bauman, W. C., 192, *212*
Bautz, E. K. F., 83, 104, *107*
Baxter, J. W. M., 160, *162*, *165*, 167
Bayer, E., 116, 122, 124, 151, 154, 158, 159, 160, 161, *162*
Beagle, R. J., 18, *74*
Begg, G., 171, 173, 179, 182, *188*
Belleau, B., 129, *162*
Bender, M. L., 34, 60, *73*, *78*
Ben-Ishai, D., 121, *162*
Benjamini, E., 161, *162*, *163*, *165*, *168*, *169*
Benson, J. V., 210, *212*, *213*
Berger, A., 67, *73*, 99, *107*, 121, 161, *162*, *164*
Berman, J. D., 99, *107*
Bernfeld, P., 5, 6, 13, 14, 24, *73*
Beyerman, H. C., 119, 122, 124, 155, 159, 160, *163*, *165*, 173, *187*
Bieber, R. E., 5, 6, 13, 24, *73*
Bier, M., 14, 29, 77
Bilibin, A. Y., 151, *169*
Billiet, H. A., 119, *163*, 173, *187*
Birr, C., 115, 129, 143, 146, 160, *169*
Blake, J., 161, *163*
Blanken, R. M., 18, *74*
Blasi, F., *107*
Bloemendol, H., 99, *108*
Blout, E. R., 146, 147, 151, *165*

Blow, D. M., 34, 77
Blumberg, S., 10, 11, 12, 19, 20, 24, 25, 27, 28, 29, 67, 73, 75, 99, *107*
Blumenthal, R., 57, *73*
Bodanszky, M., 114, 119, 120, 122, 123, 124, 137, *163*
Boers-Boonekamp, C. A. M., 122, 160, *163*
Boissonas, R. A., 121, *163, 164*
Bondi, E., 127, *163*
Bonner, A. G., 173, 174, *188*
Bortnick, N., 116, 154, *165*
Bossinger, C. D., 159, *164*
Boundy, R. H., 191, 192, 200, *212*
Bovey, F. A., 190, *212*
Boyd, G. E., 208, *212*
Boyer, R. F., 191, 192, 200, *212*
Brandenberg, D., 112, *166*
Brandenberger, H., 5, 10, *73*
Bremer, H., 112, *166*
Brey, E., 17, 18, 48, 66, *74*
Briles, D., 102, *109*
Brinkoff, O., 112, *166*
Brocklehurst, K., 7, 15, 24, 28, 29, 30, 35, 36, 37, *78*
Bronzert, T. G., 175, *188*
Brooke, G. S., 121, 136, 138, 151, *168*
Broun, G., 4, 24, 73, 77
Brown, F. S., 10, 15, 68, *78*
Brown, H. D., 5, 6, 7, 8, 24, *73, 77*
Brown, S. R., 15, 77
Brunfeldt, K., 155, 159, 161, *163, 164*
Brüning, W., 151, *162*
Bueche, F. B., 191, *213*
Bull, H. B., 5, *73*
Bumpus, F. M., 160, *165, 167*
Bünnig, K., 5, 6, 13, 14, 24, *78*
Burr, B., 102, *109*
Busch, A. W., 42, *73*
Bushuk, W., 99, *107*

C

Cahnmann, H. J., 101, *109*
Camble, R., 156, 157, *163*
Camp, P., 3, 5, *76*
Campbell, D. H., 18, 73, 79, *107, 108, 109*

Caplan, S. R., 3, 22, 24, 25, 43, 44, 45, 46, 47, 48, 49, 51, 57, 60, *73, 75*
Carpenter, F. H., 173, *188*
Case, L. C., 191, *212*
Catt, K., 151, 154, *168*
Cebra, J. J., 10, 63, 64, *73, 75*
Centeno, E. R., 161, *164*
Chan, W. C., 99, *107*
Chance, B., 58, *78*
Chang, T. M. S., 6, 14, 72, *73, 74*
Chang, W. C., 112, *165*
Change, S. P., 193, *213*
Chapman, T. M., 134, 146, 147, 151, *165*
Charlesby, A., 190, *212*
Chattopadhyay, S. K., 6, *73*
Chaturvedi, N. C., 160, *165*
Chen, C. C., 112, *165*
Chen, C. H., *187*
Cheng, L. L., 112, *165*
Chew, L., 122, 124, *164*
Chi, A. H., 112, *165*
Chibata, I., 3, 4, 72, *74, 78*
Chillemi, F., 161, *163*
Choppin, G. R., 209, *212*
Chu, B., 210, *212*
Chu, S. Q., 112, *165*
Chua, G. K., 99, *107*
Close, J., 155, *165*
Close, V. A., 122, 124, *164, 167*
Cohen, C., 11, 28, 65, *74, 76*
Colescott, R. L., 159, *164*
Cook, P. I., 59, *164*
Corigliano, M. A., *166*
Corley, L., 121, 161, *162*
Cormick, J., *212*
Cozzarelli, N. R., 83, 105, *107*
Cragg, L. H., 191, *213*
Craven, G. R., 12, *74*
Cresswell, P., 10, *74*
Crook, E. M., 7, 8, 9, 11, 15, 17, 23, 24, 28, 29, 30, 35, 36, 37, 39, 41, 42, 65, *75, 76, 78*
Cruickshank, J., 156, *168*
Crutchfield, G., 8, 11, *75*
Cuatrecasas, P., 67, *74*, 79, 81, 83, 84, 85, 86, 87, 88, 89, 90, 91, 92, 93, 95, 96, 97, 98, 99, 100, 101, 102, 103, 104, 105, 106, *107, 108, 109*
Cushman, S. W., 161, *165*

Author Index 217

D

Dale, E. C., 10, 15, 68, *78*
Das, J., 5, 6, *75*
Davie, J. M., 106, *108*
Davies, C. W., *212*
Davis, R. V., 18, *74*
Day, R. A., 23, *75*
Deer, A., 158, *163*, *167*
Degani, Y., 5, *74*
De Graaf, M. J. M., 58, *77*
DeGroot, S. R., 51, *74*
De Leer, E. W. B., 119, 124, *163*
DeLucca, M., 7, *74*, 99, *108*
Denburg, J., 7, *74*, 99, *108*
Denkewalter, R. G., *163*
Determan, H., 5, 6, 13, 15, 24, *78*
Deutsh, D. G., 99, *108*
Diamond, R. M., 210, *212*
Dietrich, H., *168*
Dijkstra, A., 119, *163*, 173, *187*
Dinius, R. H., 209, *212*
Dintzis, H. M., 12, 20, 21, *75*, 84, 91, *108*, *187*, *188*
Dirvianskyte, N., *169*
Diven, W. F., 99, *109*
Dixon, M., 34, *74*
Dobson, G. R., *212*
Doolittle, R., 173, 175, 179, *188*
Dorer, F. E., 161, *168*
Dorman, L. C., 120, 159, *163*
Doscher, M. S., 58, *74*
Dowling, L. M., 173, 174, 175, 179, 181, 183, 184, *187*
Drizlikh, G. I., 86, *108*
Drobnica, L., 185, *187*
Droscher, M. S., *74*
Du, Y. C., 112, *165*
Dubey, J. P., 195, *213*
Dunkel, W., *167*
Dunn, F. W., *168*
Du Vigneaud, V., 112, 122, 124, 160, *163*, *168*

E

Eastlake, A., 121, 161, *162*, *166*
Ebihara, H., *163*
Eckstein, H., 151, *162*

Edman, P., 171, 173, 179, 182, *188*
Edwards, B. A., 7, 15, *78*
Eichorn, J., 192, *212*
Eisele, K., 161, *169*
Eldjarn, L., 99, *108*
Ellenbroek, B. W. J., 130, 131, 151, *168*
Emerson, M. T., *212*
Endo, N., 4, *74*
Engel, A., 11, *74*
Engel, K., 86, *109*
Epand, R. F., *163*
Epstein, C. J., 11, *74*
Epton, R., 7, *73*
Erdös, E. G., 7, *77*
Erhan, S. L., 83, 105, *108*
Ericsson, L. H., 171, 173, *188*
Erlanger, B. F., 10, *74*
Ernbäck, S., 8, 17, 24, *73*, *77*, 87, *107*, *109*, *162*, *167*
Esko, K., 158, *163*
Estermann, E. F., 4, *76*
Evans, W. H., 105, 106, *108*
Evers, J., 15, *76*, 86, *109*

F

Falb, R. D., 6, 9, *76*
Falla, F., *167*
Felix, A. M., 140, 141, 151, *163*
Felsenfeld, G., *168*
Ferreira, S. H., 160, *164*
Ferriere, N., 112, *163*
Filippuson, H., 9, 12, *75*
Fink, E., 11, 17, 18, 48, 66, *74*
Flanigan, E., 124, 126, 127, 128, 130, 146, 151, 159, *163*, *166*
Flory, P. J., 191, 192, *212*
Flor, F., 160, *169*
Fohles, J., *165*
Fondy, T. P., 88, 97, 98, *108*
Förster, H. J., 9, 10, *76*
Fraefel, W., 160, *163*, *168*
Fraenkel-Conrat, H., 14, *74*
Frankel, M., 148, *169*
Franklin, E. C., 82, *108*
Freeman, D. H., 193, 196, 208, 210, *213*
Frenkel, J. K., 195, *213*
Fridkin, M., 115, 134, 143, 144, 145, 146, 148, 151, 155, 160, *163*, *164*, *167*

Frimmer, M., *165*
Fritz, H., 8, 11, 17, 18, 48, 66, *74*
Fujino, M., 121, *167*
Fukuda, K., *164*
Fuse, N., 3, 4, 72, *78*

G

Gadgary, S. D., 191, *213*
Gabel, D., 10, 12, *74*
Gafurova, N. D., *168*
Garen, A., 60, 61, *74*
Garmaise, D. L., 180, *188*
Garner, R., 156, 157, *163*, *164*
Gartner, K., 209, *213*
Garson, L. R., 118, 151, 154, *164*
Gebhardt, B. M., 11, 17, 18, 48, 66, *74*
Gervasi, J. A., 191, *213*
Gilham, P. T., 83, 86, 105, *108*
Gillert, K. E., 86, *109*
Ginsburg, A., 81, 103, *108*
Ginzburg, B. Z., 62, *74*
Gisin, B. F., 161, *163*
Givas, Sister, J., 161, *164*
Givol, D., 10, 63, 64, 67, 73, *74*, 99, *108*
Glassmeyer, C. K., 12, *74*
Glueckhauf, E., 209, *213*
Goetzl, E. J., 102, *108*
Goldberger, R. F., 105, *107*
Goldfeld, M. G., 3, 4, *74*
Goldhaber, P., 104, *108*
Goldman, R., 3, 4, 5, 22, 24, 25, 27, 39, 43, 44, 45, 46, 47, 48, 49, 50, 51, 52, 54, 56, 57, 59, 60, 61, 62, 63, *74*, *75*
Goldman, R., 189, 212, *213*
Goldring, L. S., 209, *213*
Goldstein, L., 2, 8, 10, 11, 12, 14, 15, 16, 17, 18, 19, 20, 22, 23, 25, 27, 28, 29, 30, 31, 32, 33, 34, 35, 39, 65, *74*, *75*, 76, 86, *108*, 189, 212, *213*
Good, I. J., 191, *213*
Goodman, M., 156, *164*
Goodson, L. H., 5, 14, 24, 68, *73*
Gorbunov, V. I., 160, *168*
Gordon, M., 191, 209, *212*, *213*
Gordon, S., 112, *163*
Gorecki, M., 67, *74*, 98, 99, *108*, *109*
Goring, D. A., 190, *213*
Gosnell, A. B., 191, *213*

Grahl-Nielsen, O., *164*, *168*
Gray, W. R., 181, *188*
Green, B., 151, 154, *164*
Green, J. G., 210, *213*
Green, M. L., 8, 11, *75*
Greene, L. J., *164*
Gregor, H. P., 45, *75*, 209, *213*
Gregor, L. V., 192, *213*
Grenzer, W., *165*
Gribnau, A. A. M., 99, *108*
Grieser, G., *169*
Griessbach, R., 209, *213*
Growitz, F., 22, *78*
Grubhofer, H., 7, 9, 10, 18, *75*
Grubhofer, N., *213*
Guilbault, G. G., 5, 6, 14, 24, 68, 71, 73, *75*
Günzel, G., 9, 13, *76*, 168, *188*
Gupta, S. K., *166*
Gupta, V., 12, *74*
Gurvich, A. E., 86, *108*
Gurvich, P. E., 18, *75*
Gut, V., 119, 152, 153, 158, *164*, *167*
Gutfreund, H., 34, 61, *75*, *78*
Gutman, M., 11, *75*, 77
Gutte, B., 121, 122, 151, 161, *164*
Guttman, S., 121, *163*, *164*
Gyenes, L., 86, *108*

H

Haake, E., 147, 151, *165*
Habeeb, A. F. S. A., 13, 22, *75*
Haber, E., *167*, 177, 182, *188*
Hägele, K., 151, *162*
Hagermaier, H., 122, 124, 151, 158, 159, 160, 161, *162*
Haimovich, J., 86, *109*
Halasz, I., 116, 151, 154, *162*, *164*
Hale, D. K., *213*
Hall, B. D., 104, *109*
Hall, C., 11, 65, *74*
Halpern, B., 121, 122, 124, *164*, *167*
Halstrom, J., 112, 155, 160, 161, *163*, *164*, *165*
Hamilton, P. B., 210, *213*
Hardy, P. M., 130, *164*
Harrison, I. T., 122, *164*
Harrison, S., 122, *164*

Hartman, F. C., 13, *75*
Haugland, R. P., 122, 124, *168*
Havinga, E., *169*
Hayashi, M., 102, *107*
Haynes, R., 3, 5, 22, *75*
Heilbron, E., 8, *73*
Heinz, E., 51, *75*
Helfferich, F., 26, 27, 58, 62, *75*
Helminiak, T. E., 191, *213*
Henderson, R., 34, *77*
Henkel, R., 173, *188*
Hennig, S. B., 81, 103, *108*
Herman, J., *167*
Hermodson, M. A., 171, 173, *188*
Hersh, L. S., 9, 12, 78, 84, *109*
Herzig, D. J., 23, *75*
Herzog, K. H., *169*
Hicks, G. P., 6, 13, 14, 68, 69, 70, *75*, *78*
Hindriks, H., 119, 124, 155, *163*, *165*
Hirschmann, R. F., *163*
Hiskey, R. G., 136, *164*
Hjerten, S., 84, *108*
Hoare, D. G., 179, *188*
Hoave, D. G., 15, *75*
Hochstrasser, K., 17, 18, 48, 66, *74*
Hoffer, E., 45, *75*
Hoffman, K., *164*
Hofsten, B. V., 10, *74*
Hollinden, C. S., 131, 132, 151, *168*
Holohan, P. D., 88, 97, 98, *108*
Holt, B. D., 83, 104, *107*
Hornby, W. E., 8, 9, 11, 12, 15, 24, 28, 30, 39, 40, 41, 42, 65, *75*, *76*
Hornel, S., *169*
Horner, L., 196, 197, *213*
Hornle, S., 161, *164*, *169*
Horvath, C. G., 116, 154, *164*
Hoshida, M., 11, *76*
Howe, C. W., 102, *108*
Hrabankova, E., 71, *75*
Hruby, V. J., 122, *168*
Hsing, C. Y., 112, *165*
Hsu, C. J., 8, 12, *78*
Hzu, J. Z., 112, *165*
Hu, S. C., 112, *165*
Huang, W. T., 112, *165*
Huggins, C. G., 160, *168*
Hummel, J. P., 5, *75*
Hurst, M. W., 160, *168*
Hussain, Q. Z., 11, *75*

Hutton, J. J., 161, *164*
Hutzel, M., 8, 17, 66, *74*

I

Illiano, G., *107*
Illse, D., 173, 182, *188*
Ingwall, R. T., *164*
Inman, J. K., 12, 20, 21, *75*, 84, 91, *108*, *188*
Inoue, S., 121, *167*
Intveld, R. A., 160, *163*
Inukai, N., 120, 121, 124, 126, 127, 128, 129, 133, 151, 155, *164*, *167*
Isambert, M. F., 10, *74*
Iselin, B., 119, 120, 121, 122, 124, *164*, *168*
Isliker, H. C., 86, *108*
Ives, D. A. J., 124, *164*

Jacobs, P. M., 127, 146, *164*
Jagendorf, A. T., 86, *108*
Jansen, E. F., 13, 22, *75*
Janson, J. C., 8, 17, 29, *73*
Jaquenoud, P. A., *163*
Jaquet, H. C., 63, 64, *75*
Jellum, E., 99, *108*, 122, 124, *167*
Jenssen, T. A., 7, 11, *77*
Jerina, D. M., 79, 81, 87, *108*, 139, 151, 154, 159, *165*
Jernberg, N., 155, *166*, 174, *188*
Jeschkeit, H., 173, *188*
Jiang, R. O., 112, *165*
Johnson, B. J., 127, 146, 161, *164*
Johnson, J. H., 118, 133, 134, 146, 151, 155, *169*
Johnson, J. J., 123, *167*
Johnson, M. J., 42, *75*
Jolles, J., 161, *164*
Jolles, P., 161, *164*
Jones, R. T., 210, *212*
Jorgensen, E. C., 160, *164*
Jovin, T. M., 81, 83, 105, *107*, *108*
Jung, G., 116, 151, 158, 159, 161, *162*

K

Kahn, J. R., 161, *168*
Kaiser, E., 159, *164*
Kakiuchi, K., 121, *167*
Kallos, J., 58, *75*
Kamen, M. D., 122, 161, *167*
Kapner, R. B., 86, *108*
Karlin, A., 50, 51, *77*
Karlsson, S., 158, *163*
Katchalsky, A., *213*
Katchalski, E., 2, 3, 4, 8, 10, 11, 12, 13, 14, 15, 16, 17, 18, 22, 23, 24, 25, 27, 28, 29, 30, 31, 32, 33, 34, 35, 39, 40, 43, 44, 45, 46, 47, 48, 49, 50, 51, 52, 54, 56, 57, 59, 60, 61, 62, 63, 64, 65, 68, *73*, *74*, *75*, *76*, *77*, 79, 82, 86, *108*, *109*, 115, 134, 143, 144, 145, 146, 148, 151, 160, *163*, *164*, *167*, 189, 212, *213*
Katchalsky, A., 62, *74*
Kato, I., 99, *108*, *164*
Katsoyannis, P. G., 112, *163*, *164*
Kay, G., 8, 9, 10, 11, 17, 23, 24, 42, 43, 66, *75*, *76*, *77*, *78*
Ke, L. T., 112, *165*
Kedem, O., 3, 22, 24, 25, 27, 43, 44, 45, 46, 47, 51, 52, 54, 56, 57, 59, 60, *73*, *74*, *75*
Kelleher, G., 8, 11, *78*
Kent, L. H., 86, *108*
Kerling, K. E. T., *169*
Kessler, W., 119, 121, 122, 124, *164*
Kettman, J. R., 161, *165*
Kézdy, F. J., 34, *73*
Khan, N. H., *164*
Khosla, M. C., 160, *165*
Kikuchi, Y., 121, *167*
King, M. V., 13, *76*
Kirimura, J., 6, 8, *76*
Kiryushkin, A. A., 116, 151, 154, 160, *165*, *167*, *168*
Kishida, Y., 121, 123, 133, *163*, *167*
Kivity, S., 148, *169*
Kline, G. M., 190, *213*
Klostermeyer, H., 112, 160, *164*, *165*, *166*
Knorre, D. G., 156, *165*
Knowles, C. O., 5, 24, *77*
Kobamoto, N., 3, 5, *76*
Kocy, O., *167*
Koenig, W. A., 151, 160, *162*
Kohler, P., 161, *169*

Kominz, D. R., 10, 66, *76*
Kornberg, A., 81, 83, 105, *107*, *108*
Kornet, M. J., 115, 139, 140, 151, 154, 159, *165*
Koshland, D. E., Jr., 15, *75*, 179, *188*
Kostrzewa, M., *167*
Kovacs, K., 161, *164*
Kozhevnikova, I. V., 116, 151, 154, 160, *165*, *167*, *168*
Kramer, D. N., 5, 14, 24, 68, *73*, *75*
Kraus, K. A., 209, 210, *213*
Kressman, T. R. E., 116, 154, *166*
Krivtsov, V. F., 177, *188*
Krumdieck, C. L., 121, 141, 158, 160, *165*
Kucera, G., 11, 65, *74*
Kung, Y. T., 112, *165*
Kunin, R., 116, *165*
Kuquya, T., 102, *107*
Kurihara, M., 161, *167*
Kurtz, J., 161, *164*
Kusch, P., 121, 122, *165*
Kuzovlena, O. B., *108*

L

Laidler, K. J., 63, *78*
Landman, A., 98, *109*
Lapresle, C., 86, *109*
Larsson, P. O., 6, *76*
Laufer, D. A., 134, 146, 147, 151, *165*
Laursen, R. A., 173, 174, 175, 177, 186, *188*
Lawrenson, I. J., 209, *213*
Lazare, L., 209, *213*
Leach, F. R., 83, 105, *108*
Leclerc, J., *167*
Lee, T. C., 160, *164*
Lehman, H., 173, *188*
Lenard, J., 121, 122, *165*
Leng, M., *168*
Lenhof, H., 5, 25, 45, *74*
Lentz, K. E., 161, *168*
Lerman, L. S., 18, *73*, 83, 86, 98, *107*, *108*
Letsinger, R. L., 114, 115, 139, 140, 151, 154, 159, *165*, 189, *213*
Leung, C. Y., 161, *162*, *163*, *168*, *169*

Leung, D. Y. K., 161, *165*
Leuscher, E., 18, *73*, 86, *107*
Leuschner, F., *76*
Levin, G., *166*
Levin, Y., 7, 8, 10, 11, 12, 15, 17, 19, 20, 23, 24, 25, 27, 28, 29, 30, 32, 33, 34, 35, 65, *75*, *76*, 77, 86, *108*
Levine, M., 161, *168*
Levinthal, C., 60, 61, *74*
Levy, D., 173, *188*
Lewis, F. M., 192, *213*
Lewis, M., 45, *76*
Li, C. H., 112, 161, *163*, *165*
Li, H. S., 112, *165*
Liener, I. E., 127, 128, 146, *165*
Lilly, M. D., 7, 8, 9, 10, 11, 15, 17, 23, 24, 28, 30, 39, 40, 41, 42, 43, 65, *75*, *76*, *77*, *78*
Lindenmann, A., *164*
Lindsey, A. S., 189, 212, *213*
Lingens, F., 99, *109*
Lipman, L. N., 64, *76*, *77*
Lipsky, S. R., 116, 154, *164*
Littau, V., 116, 152, *166*
Loffet, A., 122, 123, 155, *165*
Löfroth, G., 3, 5, *76*
Loh, T. P., 112, *165*
Lombardo, M., 122, 124, *165*
Lopiekes, D. V., 8, 15, *77*
Lorenz, P., 124, 125, 130, 151, *165*
Losse, G., 124, 125, 130, 151, *165*
Lotan, N., *164*
Love, J., 120, *163*
Lowey, S., 28, 65, 66, *76*, *77*
Lozier, R., *168*
Lübke, K., 139, *168*
Luck, S. M., 28, 65, *76*
Lunkenheimer, W., *165*
Lutz, F., *165*
Lynn, J., 6, 9, *76*

M

Massen van den Brink-Zimmermannova, H., 122, 160, *163*
Maat, L., 119, *163*, 173, *187*
McCauley, D. J., 208, *213*
McCormick, D. B., 83, 98, 101, 105, *107*, *108*, *109*
McCormick, H. W., 190, 210, *213*
McDonald, A., 9, *75*
MacDonnell, P. C., 5, 6, 13, 24, *73*
McIlroy, D. K., 51, *73*
McIntosh, F. C., 14, 72, *74*
McKay, R. F., 180, *188*
McLaren, A. D., 4, *76*
McLaren, J. V., 7, *73*
McLean, A., 211, *213*
Madlung, C., 124, 125, 130, 151, *165*
Maeda, H., 3, 5, *78*
Magee, M. Z., *164*
Mahadevan, V., 139, 151, 154, 159, *165*, 189, *213*
Mahajan, K., 88, 97, 98, *108*
Malek, G., 129, *162*
Malley, A., 86, *108*
Manecke, G., 2, 9, 10, 13, 14, 22, 24, *76*, 82, 86, *109*, 147, 151, *165*, 168, *188*
Manfrey, P. S., 13, *76*
Manning, J. M., 161, *165*
Manning, M., 121, 122, 124, 160, 161, *162*, *164*, *165*, *167*
Manoylov, S. E., 7, 8, 11, 17, 18, 23, *78*
Manson, J. A., 191, *213*
Mansveld, G. W. H. A., 155, *165*
Mardashev, S. R., 3, 4, *76*
Marglin, A., 119, 122, 161, *164*, *165*, *166*
Mark, H. F., 190, *212*
Marlborough, D. R., 134, 146, 147, 151, *165*
Marr, W. E., 116, 154, *166*, 193, *213*
Marshall, D. L., 127, 128, 146, *165*
Marshall, G. R., 122, 123, 124, 126, 127, 128, 130, 138, 146, 149, 151, 158, 159, 160, 161, *163*, *165*, *166*, *167*, 206, *213*
Marshall, J. S., 99, 101, *109*
Mason, S. G., 14, 72, *74*
Matthews, B. W., 34, *77*
Mattiason, B., 9, 38, 39, *76*
Maurer, P. H., 14, *77*
May, W. P., 161, *164*
Mayer, H., 133, *169*
Mayo, F. R., 192, *213*
Mazur, P., 51, *74*
Medzihradszky, K., 156, *166*
Meienhofer, H., 22, *78*
Meienhofer, J., 112, 124, 160, *166*
Meitzner, E., 116, 154, *165*
Melechen, M. E., 83, 105, *107*

Merrifield, R. B., 112, 116, 118, 119, 121, 122, 124, 137, 140, 141, 149, 151, 152, 153, 154, 155, 156, 158, 159, 160, 161, *163*, *164*, *165*, *166*, *168*, *169*, 174, *188* 206, *213*
Mertz, E. T., 99, *108*
Metzger, H., 102, *108*
Meyerson, K., 191, *212*
Michaeli, D., 161, *165*
Micheel, F., 15, 76, 86, *109*
Michelson, A. M., 10, 74
Millar, J. R., 116, 154, *166*, 193, *213*
Miller, A. W., 130, *166*
Miller, J. D., 88, 97, 98, 107, *109*
Miller, N. L., 195, *213*
Miron, T., 5, 74
Mitchell, A. R., *166*
Mitchell, E. R., 10, 66, 76
Mitchell, P., 51, 76
Mitz, M. A., 3, 4, 5, 8, 11, 15, 28, 65, 76, 86, *109*
Miura, Y., 155, 161, *166*, *168*
Mizoguchi, T., 133, 151, 160, 161, *166*, *168*
Mizrahi, R., 10, 77
Mross, G. R., 173, 175, 179, *188*
Money, C., 8, 9, 11, 76
Montalvo, J. G., 71, 75
Moore, S., *165*
Moore, G. E., 209, 210, *213*
Mori, T., 3, 4, 72, 78
Morisset, R., 102, *108*
Morton, M., 191, *213*
Mosbach, K., 5, 6, 9, 12, 38, 39, 76
Mosbach, R., 6, 76
Moudgal, N. R., 86, *109*
Munson, P. L., 104, *107*
Murakami, M., 120, 124, 126, 127, 128, 129, 133, 151, 155, *164*
Mussleman, K. E., 193, *213*
Myrin, P. A., 8, 17, 29, 73

N

Nagasawa, M., 209, *213*
Nagelschneider, G., 161, *169*
Najjar, V. A., 121, *166*
Nakamizo, N., *167*
Nakano, K., 120, 124, 126, 127, 128, 129, 133, 151, 155, *164*

Nanzyo, N., 161, *166*
Neary, J. T., 99, *109*
Nefkins, G. H. L., 133, *166*
Nelson, F., 209, *213*
Nernst, W. Z., 27, 62, 76
Neubert, K., *165*
Neudecker, M., 8, 11, 17, 66, 74
Newcomb, T. F., 11, 75, 76
Ney, K. H., 161, *166*, *167*
Nezlin, R. S., 86, *108*, *109*
Niall, H. D., 151, 154, *168*, 171, 179, *188*
Nihei, T., 10, 66, 76
Nikolayev, A. Y., 3, 4, 76
Nishimura, O., 161, *167*
Nishizaw, R., 121, 133, *167*
Nisonoff, A., 64, 76, 77
Nitecki, D. E., 121, *164*
Niu, C. I., 112, *165*
Nord, F. F., 29, 77
Northrup, L. G., 83, 105, *108*
Nyggard, A. P., 104, *109*

O

Obermeier, R., 158, *169*
Offord, R. E., *162*
Ogata, K., 13, 22, 77
Ogle, J. D., 12, 74
Ohms, J. I., 210, *213*
Ohno, M., 121, 122, 124, 161, *162*, *166*
Okada, M., 121, 123, *167*
Okuda, T., 112, 159, 161, *166*, *169*
Olafson, R. A., 133, *169*
Olson, A. C., 13, 22, 75
Olson, N. F., 13, 73
Omenn, G. S., 102, *109*, *167*
Ondetti, M. A., 114, *163*, *167*
Ong, E. B., 28, 65, 77
Ontjes, D. A., 99, 101, 102, *109*, 121, 161, *162*, *166*, *167*
Otteson, M., 13, 22, 77
Ovchinnikov, Y. A., (Yu. A.,) 116, 151, 154, 160, *165*, *167*, *168*
Ozawa, Y., 3, 5, 78

P

Packham, D. I., *213*
Paganou, A., *168*

Pages, R. A., 101, *109*
Paigen, K., 98, *109*
Paisley, H. M., 193, *213*
Park, W. K., 160, *167*
Parker, D. C., 102, *109*
Pascoe, E., 13, 24, *78*
Patchornik, A., 12, 77, 86, *108*, 115, 134, 143, 144, 145, 146, 148, 151, 155, 160, *163*, *167*
Patel, A. B., 5, 6, 7, 8, 24, *73*, 77
Patel, R. P., 13, 15, 77
Patel, V. C., 194, *213*
Patterson, J. A., 210, *212*, 213
Patton, W., 122, 124, 160, *164*, *167*
Paul, W. E., 106, *108*
Pauling, L., 86, *109*
Pecht, M., 8, 10, 11, 12, 15, 19, 20, 23, 24, 25, 27, 28, 29, 65, *75*, *76*, 86, *108*
Pejaudier, L., 11, 77
Pennington, S. N., 5, 7, 8, 14, 24, 77
Pensky, J., 99, 101, *109*
Penzer, G. R., 13, 77
Pepper, K. W., 193, *213*
Pereira, W., 122, 124, *167*
Perlmann, G. E., 28, 65, 77
Peterson, E. A., 105, 106, *108*
Peterson, G. H., 4, *76*
Pettee, J. M., 121, 136, 138, 151, *168*
Piasio, R., 122, 124, *165*
Pietta, P. G., 138, 151, 160, *167*
Pisano, J. J., 175, *188*
Pluscec, J., *167*
Pohl, H., 196, 197, *213*
Pollard, H., 88, 90, 93, 96, 98, *109*
Poltorak, O. M., 3, 4, 5, *74*, 77, *78*
Polzhofer, K. P., 161, *166*, *167*
Poonian, M. S., 105, *109*
Porath, J., 8, 12, 16, 17, 24, *73*, *74*, 77, 84, 87, *107*, *109*, 158, *162*, *163*, *167*, 187, *187*
Porter, R. R., 64, 77, 82, 86, *109*
Potts, J. T., Jr., 171, 179, *188*
Pourchot, L. M., 123, *167*
Preiss, B. A., 116, 154, *164*
Press, E. M., 82, *109*
Pressman, D., 86, *109*
Preston, J., 156, *168*
Price, S., 8, 13, 15, 77
Proath, J., *109*
Puca, G. A., 88, 95, *107*

Q

Quiocho, F. A., 13, 22, 58, 77

R

Racky, W., 157, *169*
Radda, G. K., 13, 77
Radoczy, J., 156, *166*
Ragnarsson, U., 133, *169*
Rees, A. W., 22, *75*
Reichenberg, G., 208, 209, *213*
Reilly, E., 104, *107*
Rembges, H., *165*
Ressler, C., 112, *163*
Rhener, J., Jr., 191, 192, *212*
Rice, S. A., *213*
Rice, S. N., 209, *213*
Rich, A., 83, 104, *107*
Richards, F. F., *167*
Richards, F. M., 13, 22, 58, *74*, 77
Richardson, T., 13, *73*
Riesel, E., 12, 24, 68, 77
Rimon, A., 11, 17, 28, 64, 65, *73*, *75*, 77
Rimon, S., 11, 77
Riordan, J. F., 14, 29, *78*
Roberts, C. W., 112, *163*
Robbins, J., 101, *109*
Robbins, J. B., 86, *109*
Robinson, A. B., *168*
Robinson, A. B., 121, 122, 155, 161, *165*, *167*
Robinson, W. E., 83, 105, *108*
Roeloffs, J., *169*
Roepstorff, P., 155, 159, *163*
Roeske, R. W., *166*
Rombauts, W. A., 161, *169*
Rose, B., 86, *108*
Rothe, M., *167*
Rudd, J. F., 190, *213*
Rudinger, J., 119, 152, 153, 158, *164*, *167*
Rydon, H. N., 130, *164*
Ryle, A. P., *167*

S

Sair, R. A., 13, *73*
Sakai, R., 121, *167*

Sakakibara, S., 121, 123, 133, *167*
Salvatore, G., 99, *109*
Sander, E. G., 83, *109*
Sanderson, A. R., 10, *74*
Sanger, F., 112, *167*
Sano, S., 161, *166*, *167*
Sano, Y., 124, 160, *166*
Sauer, R., 171, 179, *188*
Savery, A., 104, *107*
Sawyer, W. H., 160, *162*, *165*, *167*
Schafer, D. J., *164*
Schechter, I., 13, 67, *73*, 99, *107*
Schejter, A., 77
Schellenberger, A., 173, *188*
Scheraga, H. A., *163*, *164*
Schick, H. F., 23, 77
Schlabach, A. J., 105, *109*
Schleith, L. Z., 7, 9, 10, 18, *75*
Schleuter, R. J., 4, 5, *76*
Schlossman, S. F., *167*, *169*
Schnabel, E., 112, *166*
Schneider, F., 161, *169*
Schneider, H., *167*
Schoellmann, G., 160, *168*
Schoellmann, V. G., 161, *168*
Schoenmakers, J. G. G., 99, *108*
Schramm, W., 11, 17, 18, 48, 66, *74*
Schreiber, J., 120, *168*
Schröder, E., 139, *168*
Schroeder, W. A., 171, 184, *188*
Schult, H., 8, 11, 17, 66, *74*
Schultz, J., 186, *188*
Schwartz, R., 180, *188*
Schwyzer, R., *168*
Scott, C. D., 210, *213*
Scotchler, J., *168*
Sebastian, I., 116, 151, 154, *162*, *164*
Sehon, A. H., 2, 14, 17, 18, 68, 77, 82, 86, *108*, *109*, 161, *164*
Seki, T., 7, 11, 77
Sela, M., 10, 77, 86, *108*, *109*
Sélégny, E., 4, 24, *73*, 77
Semkin, E. P., 160, *168*
Seto, S., 155, 161, *166*, *168*
Shaltiel, Sh., 10, 77
Shapiro, J. T., *168*
Sharp, A. K., 9, 17, 23, 42, 43, *75*, *76*, 77
Shashkova, I. V., 160, *168*
Shchukina, L. A., 122, 123, 160, 161, *168*
Sheehan, J. C., 156, *168*

Sheehan, J. T., 119, 120, 122, 123, 124, 137, *163*, *167*
Shemyakin, M. M., 116, 151, 154, *165*, *168*
Shi, P. T., 112, *165*
Shigezane, K., 133, 151, *166*
Shimizu, M., 161, *162*, *163*, *168*
Shimonishi, Y., 121, 123, 161, *167*, *169*
Shin, M., 121, *167*
Shone, M. G. T., 209, *213*
Shubina, T. N., 156, *165*
Sieber, P., 121, *168*
Siegler, P. B., 34, 77
Silman, H. I., 10, 22, 43, 45, 51, 63, 64, *74*
Silman, I. H., 2, 3, 10, 11, 13, 14, 17, 18, 22, 23, 24, 25, 28, 39, 43, 44, 45, 46, 47, 48, 49, 50, 51, 60, 65, 68, *74*, 77, 79, 82, *109*
Singer, S., 9, *76*
Singer, S. J., 23, 77
Skeggs, L. T., 161, *168*
Sklyarov, L. Yu., 122, 123, 160, 161, *168*
Slade, J. H. R., 86, *108*
Slayter, H. S., 11, 28, 65, 66, *74*, *76*, 77
Sloane, R. W., Jr., *167*
Sluyterman, L. A. A. E., 58, 77, 99, *109*
Smeby, R. R., 160, *165*, *167*
Smiley, K. L., 4, *73*
Smirnova, A. P., 160, *168*
Smith, D. G., 154, *166*, 193, *213*
Smith, M. E., 194, *213*
Smith, R. L., 116, *168*
Soldano, B. A., 208, *212*
Sollner, K., 45, *75*, *76*
Somers, P. J., 7, *73*
Southard, G. L., 121, 136, 138, 151, *164*, *168*
Spink, W. W., 102, *108*
Spiro, J. G., 190, *213*
Sprossler, B., 99, *109*
Sri Ram, J., 14, 29, 77
Sroka, W., 112, *166*
Stark, G. R., 172, 173, 174, 175, 179, 181, 183, 184, *187*, *188*
Steers, E., 88, 90, 93, 96, 98, *109*
Steffen, K. D., *167*
Steinbuch, M., 11, 77
Stelakatas, G. C., *168*
Stepanov, V. M., *188*
Steuben, K. C., 156, *164*

Stevenson, K. J., 98, *109*
Stewart, F. H. C., 121, *168*
Stewart, J. M., 112, 119, 120, 121, 122, 155, 160, 161, *164, 166, 168, 169*, 174, *188*
Stirling, C. J. M., 130, *166*
Stockmayer, W. H., 191, *213*
Stone, I., 4, 77
Strandberg, G. W., 4, *73*
Stryer, L., 122, 124, *168*
Stupp, Y., 10, 11, 77
Sugihara, H., 121, *167*
Sugiyama, H., 155, *168*
Summaria, L. J., 8, 11, 15, 28, 65, *76*, 86, *109*
Sundaram, P. V., 12, 63, *78*
Sundheim, B. R., 209, *213*
Surbeck-Wegmann, E., *168*
Surinov, B. P., 7, 8, 11, 17, 18, 23, *78*
Suzuki, H., 3, 5, *78*
Suzuki, K., *164*
Svendsen, I., 13, 22, 77
Svensson, B., 13, 77
Swan, J. M., 112, *163*
Swilley, E. L., 42, *73*

T

Takahashi, M., 99, *107*
Takamura, N., 133, 151, *166*
Takashima, H., 121, 122, 124, 160, *168*
Taketomi, N., 4, *78*
Talmage, D. W., 86, *109*
Tanford, C., 35, *78*
Tang, K. L., 112, *165*
Tanimura, T., 119, *165*
Taniuchi, H., *109*
Terminiello, L., 29, 77
Tesser, G. I., 130, 131, 151, *168*
Thampi, N. S., 160, *168*
Theorell, H., 58, *78*
Theysohn, R., *167*
Thiele, E. W., 26, 58, *78*
Thomas, D., 4, 24, *73*, 77
Thomsen, J., 159, 161, *163, 164*
Thomson, J. R., 5, 68, *73*
Thompson, E. B., 88, 97, 98, 107, *109*
Thompson, R. C., 130, *164*
Thurmond, C. D., 191, *213*
Tilak, M. A., 116, 119, 131, 132, 151, *164, 168*

Tometsko, A., *164*
Tomino, A., 98, *109*
Tosa, T., 3, 4, 72, *74, 78*
Tosteson, D. C., 161, *164*
Toyama, M., 161, *166*
Trask, E. G., *164*
Tregear, G. W., 151, 154, *168*
Trentham, D. R., 61, *68*
Tripp, S. L., *168*
Tritsch, G. L., *164, 168*
Truffa-Bachi, P., 106, *109*
Tsang, Y., 28, 65, 77
Tumanova, A. E., *108*
Tweedale, A., 63, *78*

U

Undenfriend, S., 161, *164*
Updike, S. J., 6, 13, 14, 68, 69, 70, *75, 78*
Usami, A., 4, *78*
Usdin, V. R., 5, 14, 24, *72*
Uziel, M., 6, *78*

V

Vaidya, V. M., 146, 147, 151, *165*
Vallee, B. L., 14, 29, *78*
Van Amburg, G., 3, 5, *76*
vand der Ploeg, M., 13, 24, *78*
Van Doninck, A. H., 119, *163*, 173, *187*
van Duijn, P., 13, 24, *78*
Van Kraaikamp, M., 99, *108*
Van Velthuyzen, H., 119, *163*, 173, *187*
Van Zoest, W. J., 122, 160, *163*
Vasta, B. M., 5, 14, 24, *72*
Venkatachalam, T. K., 193, *213*
Vesa, V., 112, *168*
Visser, S., *169*
Vlasov, G. P., 151, *169*
von Heyl, G. C., 22, *78*
Vorobeva, E. S., 3, 4, 5, *74*, 77, *78*
Vretblad, P., 12, *74*

W

Wage, M. G., 105, 106, *108*
Wagner, T., 8, 12, *78*

Walsh, K. A., 5, 22, 75, 171, 173, *188*
Wan, R. E., 5, 6, 13, 14, 24, *73*
Wang, S. S., 134, 137, 151, 159, 161, *169*
Wang, Y., 112, *165*
Waterfield, M., 177, 182, *188*
Watkins, W. B., *164*
Watson, D. M., 6, 13, *73*
Watts, R. E., 209, *213*
Webb, E. C., 34, *74*
Webb, T., 86, *109*
Weber, U., 161, *164*, *169*
Weber, V. U., 161, *169*
Weeds, G., 28, 66, *76*
Weetall, H. H., 6, 7, 9, 10, 12, 15, 18, 20, 68, *78*, 84, *109*
Weinstein, Y., 67, *74*, 99, *108*
Weintraub, B., 102, *109*
Weissbach, A., 105, *109*
Weitzel, G., 161, *164*, *169*
Weliky, N., 10, 15, 18, 68, *78*
Wellner, D., 11, 77
Werle, E., 8, 11, 17, 18, 48, 66, *74*
Westman, T. L., 18, 65, *78*
Weygand, F., 119, 131, 133, 151, 158, *169*
Wharton, C. W., 7, 15, 24, 28, 30, 35, 36, 37, *78*
Wheeler, A., 26, 58, *78*
Wheeler, K. P., 7, 15, *78*
Whitaker, J. R., 60, *78*
Whitney, D. C., 210, *212*
Whittam, R., 7, 15, *78*
Wide, L., 84, *109*
Wiedermann, M., 8, 17, 66, *74*
Wieland, T., 5, 6, 13, 14, 24, *78*, 115, 129, 143, 146, 151, 157, 160, *169*
Wigzell, H., 105, *109*
Wijdenes, J., 99, *109*
Wilchek, M., 67, *74*, *78*, 81, 84, 85, 87, 95, 98, 99, 100, 101, 102, *107*, *108*, *109*
Wildi, B. S., 118, 133, 134, 146, 151, 155, *169*
Wiley, R. H., 193, *213*
Williams, D. A., 42, *73*
Wilson, R. J. H., 7, 9, 10, 11, 17, 23, 24, 42, 43, *75*, *76*, *78*
Windridge, G. C., 160, *164*
Winkler, C. A., 190, *213*

Winter, A., 8, *73*
Wissenbach, H., *169*
Wissler, F. C., 64, *76*, 77
Witkop, B., 161, *164*
Woernley, D. L., *76*
Woffard, C. F., 191, *213*
Wofsy, L., 102, 106, *109*
Wold, F., 13, 23, *75*, *78*
Wolman, Y., 148, *169*
Woodward, R. B., 133, *169*
Woolley, D. W., 121, 122, 160, 161, *166*, 168, *169*
Wright, L. D., 83, 105, *109*
Wuu, T. C., 160, *165*, *167*
Wyman, D. P., 191, *212*

Y

Yagi, Y., 86, *109*
Yajima, H., 161, *167*
Yanari, S. S., 4, *76*
Yang, H. Y. T., 7, 77
Yaron, A., *167*, *169*
Yie, Y. H., 112, *165*
Yoeman, G. D., *212*
Yonetani, T., 58, *78*
Yoshida, T., 6, 8, *76*
Yoshimuri, S., 121, *167*
Young, G. T., 156, 157, *163*, *164*
Young, J. D., 112, 120, 161, *162*, *163*, 168, *169*
Young, M., 99, *107*
Young, M. A., 193, *213*

Z

Zabel, R., 112, *166*
Zahn, H., 22, *78*, 112, 159, 161, *166*, *169*
Zamani, M., *167*
Zec, J., 210, *213*
Zelinski, P., 191, *213*
Zenftman, H., 211, *213*
Zervas, L., *168*
Zimm, B. H., 191, *213*
Zingaro, R. A., 6, *78*
Zittle, C. A., 3, *78*

Subject Index

A

Acetylcholinesterase, 5, 24, 50, 99
Acid phosphatase, 4
ACTH, 161
Acylase, 4, 6
Adenosine triphosphatase, 6
Affinity chromatography, 79–107
  advantages of, 81
  anchoring arms, 95–97
  binding strength effects, 80
  chemically modified proteins, separation of, 100
  conditions important for, 80, 94–95
  elution by denaturation, 94
    by inhibitors, 95
  estimation of bound material, 85
  coupling of diazotized ligands to, 90
  ligand selection for, 84
  number of linkages to bound protein, 104
  pH effects in coupling, 104
  porosity effects in, 83
  properties of supports, 83
  purification of peptides by, 101
  removal of intact protein-ligand complex, 94
  steric interference in, 85
Agarose
  alkylated derivatives of, 88
  diazonium derivative of, 88
  diazo derivative of, 88
  sulfhydryl derivatives of, 89
  as support for adsorbents, 84, 86
Alcohol dehydrogenase, 5, 24, 58
Aldolase, 5, 6
Alkaline phosphatase, 4, 6, 24, 25, 43, 45, 50, 60, 61
Alkyl esters, in solid phase peptide synthesis, 130
Amide support, in solid phase peptide synthesis, 137

Amino acids
  resins for analysis of, 210
  resolution with insoluble aminoacylase, 72
Aminoacylase, mold, insoluble resolution of amino acids with, 72
Amino acyl t-RNA synthetase, 7
Aminoethyl cellulose, linking of enzymes to, 18
Aminopolystyrene, 174, 180, 206–208, 211
Amino protection, polymeric, in peptide synthesis, 138–141
Ammonolysis, in solid phase peptide synthesis, 123
Amylase, 7
Amyloglucosidase, 7
Amyloid protein, human, 99
Anataminide, 160
Anchoring arms, see Affinity chromatography
Angiotensin, 157, 160
Angiotensinylbradykinin, 149
Antibodies
  attachment to insoluble support, 102
  purification by affinity chromatography, 102
Antigens
  attachment to cellulose with carbodiimides, 86
  insoluble derivatives of, 68
  specificity after attachment to supports, 82
Apyrase, 7
Asparaginase, 4
Aspartic acid residues, problem in solid phase peptide degradation, 177
Automation
  of peptide degradation with solid support, 176
  of protein degradation, 171
  of solid phase peptide synthesis, 155

Subject Index

Autoradiography, peptide distribution in polymer support, 153
Avidin, 98
Azide coupling, in solid-phase peptide synthesis, 141
Azides, linking of enzymes to polymers via, 15, 20

B

Benzhydryl polymeric support, in solid phase peptide synthesis, 136
1-Benzoylisopropenyl protecting group, 121, 136
Benzyl esters, modified, in solid phase peptide synthesis, 118–126
  lability of, 121
  side reactions, 118
Biphenylisopropyloxycarbonyl protecting group, 121
Bradykinin, 144, 149, 160
  potentiating factor, 160
Bromelain, 7, 24, 28, 30
Bromoacetyl polystyrene in peptide synthesis, 132, 133
α-Bromopropionyl polymer, in solid phase peptide synthesis, 133
Brush polymers, 154
t-Butyloxycarbonyl azide, amino protection in peptide degradation, 173

C

Carbodiimides
  amino attachment to agarose with, 88
  blocking of peptide carboxyl groups with, 179
  carboxyl attachment to agarose with, 88
  coupling of peptide to resins with, 175
  enzyme attachment to polymers with, 15
  polymeric, in solid phase peptide synthesis, 148
  protein attachment to cellulose with, 86
Carbonyldiimidazole
  in solid phase peptide degradation, 173
  in solid phase peptide synthesis, 120
Carboxyl groups, of side chains, specific blocking of, 179
Carboxyl protection, in solid phase peptide synthesis, 115–138
Carboxylic polymers,
  covalent binding of enzymes to, 14
Carboxypeptidase, 7, 13, 58, 98
Casein, 161
Catalase, 4, 7, 13
Cell structures, isolation by affinity chromatography, 105
Cells, isolation by affinity chromatography, 105
Cellulose
  diazo derivative of, 86
  linking of enzymes to, 17
  polynucleotide derivatives of, 105
  as support in affinity chromatography, 86
Chorionic somato-mammotropin, 102
Chloroacetyl polystyrene, in peptide synthesis, 132
Chloromethyl polystyrene, 204–206
  amination of, 206–208
  cross-linking in, 204
  Friedel–Crafts reagent, 204
  linear polymer formation in, 205
  methylene bridging, 204
  nitration of, 174
  in peptide degradation, 174, 175
  in solid phase peptide synthesis, 116
  in synthesis of polymeric alkyl esters, 131
Cholinesterase, 5, 8, 68
Chorismate mutase, 99
Chromatography, affinity, see Affinity chromatography
Chymotrypsin, 4, 5, 8, 13, 23, 24, 27, 30, 34, 43, 58, 65, 98, 161
Citrate synthetase, 5
Collagen, 161
Coupling reagents, polymeric, in peptide synthesis, 143
Creatine kinase, 9, 43
Cyanogen bromide, activation of agarose and Sephadex with, 87
Cyclic peptides
  synthesis with polymeric coupling reagents, 144
    with polymeric side-chain protection, 141
    with polymeric sulfonylphenyl esters, 146
Cytochrome c, 161

D

Deoxyribonuclease, 9
Dialkylsulfonium polystyrenes, esterification with, 120
Diamines
  for attaching proteins to agarose, 87
  reaction with chloromethyl polystyrene, 174, 175
Diastase (amylase), 4, 6, 9
Diazo groups
  linking of enzymes to polymers via, 18
  reduction of, 90, 95
2,6-Dinitrobenzene-1,4-disulfonate, 175
Diffusion in polymers
  effect on solid phase peptide synthesis, 149
  of reagents for peptide degradation, 185
N,N-Dimethylformamide-dineopentyl-acetal, in solid phase peptide syntheis, 120
DNA polymerase, 105

E

Electrostatic interactions, in enzyme-substrate reactions, 28
Eledoisin, 161
Enzyme columns, 39–43
  kinetic behavior of, 39
Enzyme deficiency, therapy with enzyme microcapsules, 72
Enzyme electrodes, 70
Enzyme membranes, 43–63
  enzyme-collodion membranes, structure of, 43
  kinetic behavior of, 51
  Michaelis constants of, 62
  pH dependence of activity, 45
  unstirred layers, effects of, 62
Enzymes, crystalline, diffusional rate-limiting effects, 58
Enzymes insoluble derivatives of, 2–23
  in biochemical analysis, 68
  chemical modification, effects of, 28
  covalent binding to insoluble carriers, 6–12, 14–22
  diffusion limitations, 26
  functional groups for covalent binding, 14
  immobilization by adsorption, 3–5
  intermolecular cross-linking, 13, 22–23
  kinetic behavior, 26–39
  Michaelis constants, 27, 34–38
  multienzyme systems, 38
  occlusion in polymeric matrices, 5–6, 13–14
  pH-activity profiles, 29–34
  pH stability, 25
  porous sheets, 43
  preparation, 2
  properties of stirred suspensions, 42
  steric effects, 27
  storage stability, 23
  thermal stability, 24
  uses, 63–72
Ethylene-maleic anhydride copolymer
  linking of enzymes to, 17
  polyionic character of, 18

F

Ferredoxin, 161
Fibrinogen, cleavage by EMA-trypsin, 64
Fibrinopeptide A, 161
Ficin, 9, 24, 28, 30, 41
Flavokinase, 98
Fructose-1,6-diphosphatase, 9

G

β-Galactosidase, 9, 43, 96, 98
Glass, covalent, linking of proteins to, 20
γ-Globulin cleavage by insoluble papain, 64
Glucagon, 161
Glucoamylase, 4
Glucosaminol as hydrophilic reactant for hydrophobic resins, 180
Glucose oxidase, 4, 6, 9, 24, 69, 70
Glucose-6-phosphate dehydrogenase, 5, 9, 25, 38
Glutamic acid residues, in solid phase peptide degradation, 177
Glutamic-pyruvic transaminase, 6
Glutamine synthetase, 103
Glyceraldehyde-3-phosphate dehydrogenase, 9
Glycerol-3-phosphate dehydrogenase, 97, 98
Gramicidin, 160

H

Haptenes, 106
Hemoglobin, 99, 161
Hexokinase, 5, 9, 38
Histidyl-tRNA, 105
Histidyl-tRNA-synthetase, 105
Hydrazides
  derivative of polyacrylamide, 91
  linking of enzymes to polymers via, 20
  of protected peptides, synthesis by solid phase method, 134
Hydrazinolysis, in solid phase peptides synthesis, 123–124
Hydrochloric acid, cleavage with, in peptide degradation, 177
Hydrogen fluoride, cleavage with, in solid phase peptide synthesis, 121
Hydroxymethyl polystyrene, in solid phase peptide synthesis, 119
Hydroxysuccinimide esters, polymeric, in solid phase peptide synthesis, 146

I

Insulin, 102, 103, 106, 161
Invertase, 5, 9, 24
Ion-exchange resins
  amino acid analysis and, 210
  cross-linking heterogeneity, 208
  swelling of, 208
  theoretical models of, 209
Isothiocyanates
  hydrolysis of, 184
  insoluble
    for attachment of proteins, 86
    for peptide degradation, 179–184
    synthesis of, 180
  in peptide degradation, 171–179
Isotope dilution, of thiohydantoins, 175

L

Lactic dehydrogenase, 6, 10, 24, 43, 69
Lipase, 5, 10
Luciferase, 10
Lymphoid cells, 106
Lysozyme, 161

M

Mechanical stability, of polymers, effect on solid phase peptide synthesis, 149
Melittin, 161
Membranes, see Enzyme membranes
  occlusion of enzymes in, 14
Meromyosins, cleavage by EMA-trypsin, 65
Methylisothiocyanate, 175
4-(Methylthio) phenyl ester, protection and activation of carboxyls with, 127–130
Michaelis constants, 27, 34–38, 62
Myoglobin, 161
Myosin, cleavage by EMA-trypsin, 65

N

Nernst diffusion layers, and enzyme membranes, 62
Nitro groups, reduction in polymers with $SnCl_2$, 174, 180, 211
Nitrophenyl esters, polymeric, in solid phase peptide synthesis, 144
Nitrophenyl sulfenyl protecting group, 121
Nucleic acids
  coupling to agarose, 105
  specific adsorbents, 104–105

O

Orsellinic decarboxylase, 6
Oxytocin, 123, 124, 130, 160

P

Papain, 5, 6, 10, 13, 23, 25, 27, 28, 30, 43 45, 47, 49, 57, 58, 60, 64, 99
Parathyroid hormone, 161
Pepsin, 10, 24
Peptide(s), purification by affinity chromatography, 101
Peptide bonds, lability in acid, 186
Peptide coupling, in solid phase peptide synthesis, 133

Peptide degradation
    effects of oxygen and aldehydes on, 173, 175, 182
    hydrophilic character of resins, for, 185
    with insoluble reagent, 179
    with solid supports, 171–187
        properties of resins for, 184
Peptide synthesis, solid phase
    analysis of products, 158
    automation, 155
    effects of physical properties of supports, 148–155
    general concepts, 112–115
    heterogeneity of reactive sites in, 153
    hydrophilic supports, 155
    soluble polymeric supports, 154
    strategy, 114
Peroxidase, 10, 68
pH, local of immobilized enzymes, 33
pH-activity profiles, 29–34
Phenyl esters in solid phase peptide synthesis, 126–130
    coupling reaction, 126
    intrachain disulfide synthesis with, 126
Phenylisothiocyanate, 172, 175
    insoluble reagent from, 179–184
Phosphoglucomutase, 5
Phosphoglyceromutase, 6
4-Picolyl esters, for protection of carboxyls, 156
Plasminogen, 99
Polistis kinin, 161
Polyacrylamide
    acyl azide, 92
    comparison with agarose for affinity chromatography, 93
    hydrazide, 91
    occlusion of enzymes with, 13
    stable cross-links for, 186
    as support for affinity chromatography, 91–93
Polyglutamyl folic acid, 160
Polymeric supports, soluble, in solid phase peptide synthesis, 154
Polystyrene, 189–212
    branching of, 190
    cross-linked, 192–195
    effects of cross-linking on properties for peptide degradation, 186
    diazo derivative of, 86
    nitration of, 180, 211
    penetration temperature, 202
    properties important for use as insoluble supports, 190
    statistical models, 191
    as support in affinity chromatography, 86
Polystyrene beads
    bead size as function of cross-linking, 193
    chain entanglement in, 193
    chloromethylation of, see Chloromethyl polystyrene
    cross-linking heterogeneity in, 193
    detailed structure as function of method of polymerization, 194
    effect of size on peptide degradation, 186
    preparation of, 195–200
        benzoyl peroxide catalyst, 196
        chain length, 199
        chain termination, 200
        divinylbenzene purity, 196
        effects of impurities, 196
        mixing rate, 196
        polymer degradation, 200
        polymerization rate, 199
        sphere size, 199
        suspending and stabilizing agents, 197
        temperature regulation, 196
    properties of ion-exchange derivatives, 208–210
    structural models of, 192
    sulfonation of, see Sulfonyl polystyrene
    surface capacity, relation to total capacity, 187
Polystyrene films and membranes, 192
Polystyrene sols and gels, 190
Pore size, of polymers, effect on solid phase peptide synthesis, 149
Pronase, 10
Protease inhibitors, adsorption to EMA-enzymes, 66
Protecting groups, side chain, removal with HF, 121
Proteinase (bacterial), 10
Prothrombin, 11
Pyridoxal kinase, 99
Pyridoxamine phosphate agarose, use in purifying ribosomes, 107
Pyruvate kinase, 11, 43

Q

Quinaldine esters, polymeric, in solid phase peptide synthesis, 147

R

Racemization in peptide synthesis, 115
 with polymeric amino protection, 138
 with polymeric coupling reagents, 144
Receptor structures, isolation by affinity chromatography, 105
Renin, 11
Rennin, 11
Resitol polymers in solid phase peptide synthesis, 125
Ribonuclease, 5, 11, 13, 58, 67, 98, 99, 102, 161
 inhibitor, 99
Ribosomes, purification by affinity chromatography, 107

S

Saponification, in solid phase peptide synthesis, 123
Sequenator, 171, 179
Serum protein
 estradiol-binding, 99
 thyroxine-binding, 99
Side-chain protection, polymeric, in solid phase peptide synthesis, 141
Sodium dithionite, for reduction of diazo derivatives, 90, 95
Solvation, of polymers, effect on solid phase peptide synthesis, 149
Staphylococcal nuclease, 95, 98, 100, 101, 161
Steric hindrance
 in affinity chromatography, 85
 in solid phase peptide synthesis, 116
Steroid Δ^1-dehydrogenase, 6
Streptokinase, 11
Styrenes, polymerization of substituted derivatives, 200
Subtilopeptidase A (subtilisin Carlsberg), 11, 23, 25, 27, 28, 34
Subtilopeptidase B (subtilisin Novo), 13, 23, 24

Succinimidomethyl support, for solid phase peptide synthesis, 134
Sulfonylphenyl esters, polymeric, in solid phase peptide synthesis, 146
Sulfonyl polystyrene, 201–204
 by chlorosulfonic acid-methylene chloride method, 203
 removal of impurities before sulfonation, 201
 sulfone cross-links in, 203
 by sulfuric acid method, 201
 by sulfuric acid-perchloroethylene method, 202
Surface area, of polymers, in solid phase peptide synthesis, 153
Swelling, of polymers
 effect of in solid phase peptide synthesis, 116, 149
 in peptide degradation, 174, 184–186
 in synthesis of resin for, 180

T

Thiazolinones, 171
 protective effect of dithioerythritol on, 173
Thiohydantoins, 171
 gas-liquid chromatography of, 175, 177
 protective effect of dithioerythritol on, 173
 radioactive, in peptide degradation, 175
 thin layer chromatography of, 175, 177
Thrombin, 11
Tobacco mosaic virus, coat protein, 161
Transesterification, in solid phase peptide synthesis, 124
Trypsin, 5. 6, 11, 13, 23, 25, 28, 30, 34, 64
 inhibitors, adsorption to EMA-trypsin, 66
Tyrosinase, 98
Tyrosine amino transferase, 97, 98, 107

U

Urease, 6, 12, 68, 71

V

Valinomycin, 161
Vasopressin, 124, 160

W

Woodward's reagent K, linking of enzymes to polymers with, 15

Z

Zymogens, activation by insoluble enzymes, 65